Roger Lee (Ed.)

Software Engineering, Artificial Intelligence, Networking
and Parallel/Distributed Computing 2011

T0189533

Studies in Computational Intelligence, Volume 368

Editor-in-Chief

Prof. Janusz Kacprzyk
Systems Research Institute
Polish Academy of Sciences
ul. Newelska 6
01-447 Warsaw
Poland
E-mail: kacprzyk@ibspan.waw.pl

Further volumes of this series can be found on our
homepage: springer.com

Vol. 346. Weisi Lin, Dacheng Tao, Janusz Kacprzyk, Zhu Li,
Ebroul Izquierdo, and Haohong Wang (Eds.)
Multimedia Analysis, Processing and Communications, 2011
ISBN 978-3-642-19550-1

Vol. 347. Sven Helmer, Alexandra Poulovassilis, and
Fatos Xhafa
Reasoning in Event-Based Distributed Systems, 2011
ISBN 978-3-642-19723-9

Vol. 348. Beniamino Murgante, Giuseppe Borruso, and
Alessandra Lapucci (Eds.)
*Geocomputation, Sustainability and Environmental
Planning,* 2011
ISBN 978-3-642-19732-1

Vol. 349. Vitor R. Carvalho
Modeling Intention in Email, 2011
ISBN 978-3-642-19955-4

Vol. 350. Thanasis Daradoumis, Santi Caballé,
Angel A. Juan, and Fatos Xhafa (Eds.)
*Technology-Enhanced Systems and Tools for Collaborative
Learning Scaffolding,* 2011
ISBN 978-3-642-19813-7

Vol. 351. Ngoc Thanh Nguyen, Bogdan Trawiński, and
Jason J. Jung (Eds.)
*New Challenges for Intelligent Information and Database
Systems,* 2011
ISBN 978-3-642-19952-3

Vol. 352. Nik Bessis and Fatos Xhafa (Eds.)
*Next Generation Data Technologies for Collective
Computational Intelligence,* 2011
ISBN 978-3-642-20343-5

Vol. 353. Igor Aizenberg
*Complex-Valued Neural Networks with Multi-Valued
Neurons,* 2011
ISBN 978-3-642-20352-7

Vol. 354. Ljupco Kocarev and Shiguo Lian (Eds.)
Chaos-Based Cryptography, 2011
ISBN 978-3-642-20541-5

Vol. 355. Yan Meng and Yaochu Jin (Eds.)
Bio-Inspired Self-Organizing Robotic Systems, 2011
ISBN 978-3-642-20759-4

Vol. 356. Slawomir Koziel and Xin-She Yang
(Eds.)
Computational Optimization, Methods and Algorithms, 2011
ISBN 978-3-642-20858-4

Vol. 357. Nadia Nedjah, Leandro Santos Coelho,
Viviana Cocco Mariani, and Luiza de Macedo Mourelle (Eds.)
*Innovative Computing Methods and their Applications to
Engineering Problems,* 2011
ISBN 978-3-642-20957-4

Vol. 358. Norbert Jankowski, Włodzisław Duch, and
Krzysztof Grąbczewski (Eds.)
Meta-Learning in Computational Intelligence, 2011
ISBN 978-3-642-20979-6

Vol. 359. Xin-She Yang, and Slawomir Koziel (Eds.)
*Computational Optimization and Applications in
Engineering and Industry,* 2011
ISBN 978-3-642-20985-7

Vol. 360. Mikhail Moshkov and Beata Zielosko
Combinatorial Machine Learning, 2011
ISBN 978-3-642-20994-9

Vol. 361. Vincenzo Pallotta, Alessandro Soro, and
Eloisa Vargiu (Eds.)
Advances in Distributed Agent-Based Retrieval Tools, 2011
ISBN 978-3-642-21383-0

Vol. 362. Pascal Bouvry, Horacio González-Vélez, and
Joanna Kolodziej (Eds.)
*Intelligent Decision Systems in Large-Scale Distributed
Environments,* 2011
ISBN 978-3-642-21270-3

Vol. 363. Kishan G. Mehrotra, Chilukuri Mohan, Jae C. Oh,
Pramod K. Varshney, and Moonis Ali (Eds.)
Developing Concepts in Applied Intelligence, 2011
ISBN 978-3-642-21331-1

Vol. 364. Roger Lee (Ed.)
Computer and Information Science, 2011
ISBN 978-3-642-21377-9

Vol. 365. Roger Lee (Ed.)
*Computers, Networks, Systems, and Industrial
Engineering 2011,* 2011
ISBN 978-3-642-21374-8

Vol. 366. Mario Köppen, Gerald Schaefer, and
Ajith Abraham (Eds.)
Intelligent Computational Optimization in Engineering, 2011
ISBN 978-3-642-21704-3

Vol. 367. Gabriel Luque and Enrique Alba
Parallel Genetic Algorithms, 2011
ISBN 978-3-642-22083-8

Vol. 368. Roger Lee (Ed.)
*Software Engineering, Artificial Intelligence, Networking and
Parallel/Distributed Computing 2011,* 2011
ISBN 978-3-642-22287-0

Roger Lee (Ed.)

Software Engineering, Artificial Intelligence, Networking and Parallel/Distributed Computing 2011

 Springer

Editor

Prof. Dr. Roger Lee
Central Michigan University
Software Engineering & Information Technology Institute
Mt. Pleasant, MI 48859
U.S.A.
E-mail: lee1ry@cmich.edu

ISBN 978-3-642-26867-0   ISBN 978-3-642-22288-7 (eBook)

DOI 10.1007/978-3-642-22288-7

Studies in Computational Intelligence   ISSN 1860-949X

Typeset & Cover Design: Scientific Publishing Services Pvt. Ltd., Chennai, India.

Printed on acid-free paper

9 8 7 6 5 4 3 2 1

springer.com

Preface

The purpose of the 12th International Conference on Software Engineering, Artificial Intelligence, Networking and Parallel/Distributed Computing (SNPD 2011) held on July 6–8 2011 Sydney, Australia was to bring together researchers and scientists, businessmen and entrepreneurs, teachers and students to discuss the numerous fields of computer science, and to share ideas and information in a meaningful way. Our conference officers selected the best 14 papers from those papers accepted for presentation at the conference in order to publish them in this volume. The papers were chosen based on review scores submitted by members of the program committee, and underwent further rounds of rigorous review.

In Chapter 1, Loretta Davidson et al. In this paper, we analyze the Environment II module of this data set using variable clustering to produce meaningful clusters related to questionnaire sections and provide information to reduce the number of demographic variables considered in further analysis. Case level clustering was attempted, but did not produce adequate results.

In Chapter 2, Xianjing Wang et al. This paper presents an approach of polar coordinate-based handwritten recognition system involving Support Vector Machines (SVM) classification methodology to achieve high recognition performance. We provide comparison and evaluation for zoning feature extraction methods applied in Polar system. The recognition results we proposed were trained and tested by using SVM with a set of 650 handwritten character images. All the input images are segmented (isolated) handwritten characters. Compared with Cartesian based handwritten recognition system, the recognition rate is more stable and improved up to 86.63%.

In Chapter 3, Hong Zhou et al. This paper provides the mathematical analysis of the statistical delay bounds of different levels of Constant Bit Rate (CBR) traffic under First Come First Served with static priority (P-FCFS) scheduling. The mathematical results are supported by the simulation studies. The statistical delay bounds are also compared with the deterministic delay bounds of several popular rate-based scheduling algorithms. It is observed that the deterministic bounds of the scheduling algorithms are much larger than the statistical bounds and are overly conservative in the design and analysis of efficient QoS support in wireless access systems.

In Chapter 4, Ming Zhang. A new learning algorithm for SS-HONN is also developed from this study. A time series data simulation and analysis system, SS-HONN Simulator, is built based on the SS-HONN models too. Test results show that every error of SS-HONN models are from 2.1767% to 4.3114%, and the average error of Polynomial Higher Order Neural Network (PHONN), Trigonometric Higher Order Neural Network (THONN), and Sigmoid polynomial Higher Order Neural Network (SPHONN) models are from 2.8128% to 4.9076%. It means that SS-HONN models are 0.1131% to 0.6586% better than PHONN, THONN, and SPHONN models.

In Chapter 5, Tsuyoshi Miyazaki et al. In this paper, we examine a method in which distinctive mouth shapes are processed using a computer. When lip-reading skill holders do lip-reading, they stare at the changes in mouth shape of a speaker. In recent years, some researches into lip-reading using information technology has been pursued. There are some researches based on the changes in mouth shape. The researchers analyze all data of the mouth shapes during an utterance, whereas lip-reading skill holders look at distinctive mouth shapes. We found that there was a high possibility for lip-reading by using the distinctive mouth shapes. To build the technique into a lip-reading system, we propose an expression method of the distinctive mouth shapes which can be processed using a computer. In this way, we acquire knowledge about the relation between Japanese phones and mouth shapes. We also propose a method to express order of the distinctive mouth shapes which are formed by a speaker.

In Chapter 6, Pinaki Sarkar et al. Resource constraint sensors of a *Wireless Sensor Network (WSN)* cannot afford the use of costly encryption techniques like *public key* while dealing with *sensitive data*. So *symmetric key encryption* techniques are preferred where it is essential to have the same cryptographic key between communicating parties. To this end, keys are preloaded into the nodes before deployment and are to be established once they get deployed in the *target area*. This entire process is called key predistribution. In this paper we propose one such scheme using *unique factorization of polynomials over Finite Fields*. To the best of our knowledge such an elegant use of Algebra is being done for the first time in WSN literature. The best part of the scheme is large number of node support with very small and uniform *key ring* per node. However the resiliency is not good. For this reason we use a special technique based on Reed Muller codes proposed recently by Sarkar, Saha and Chowdhury in 2010. The *combined scheme* has good resiliency with huge node support using very less keys per node.

In Chapter 7, Humayra Binte Ali et al. In this paper we proposed two different subspace projection methods that are the extensions of basis subspace projection methods and applied them successfully for facial expression recognition. Our first proposal is an improved principal component analysis for facial expression recognition in frontal images by using an extension of eigenspaces and we term this as WR-PCA (region based and weighted principal component analysis). Secondly we proposed locally salient Independent component analysis(LS-ICA) method for facial expression analysis. These two methods are extensively discussed in the rest of the paper. Experiments with Cohn-kanade database show that these techniques achieves an accuracy rate of 93% when using LS-ICA and 91.81% when WR-PCA and 83.05% when using normal PCA. Our main contribution here is that by performing WR-PCA, which is an extension of typical PCA and first investigated by us, we achieve a nearly similar result as LS-ICA which is a very well established technique to identify partial distortion.

In Chapter 8, Belal Chowdhury et al. The paper outlines a Health Portal model for designing a real-time health prevention system. An application of the architecture is described in the area of web Health Portal.

In Chapter 9, Mark Wallis et al. This paper presents a model of policy configuration that harnesses the power of the Internet community by presenting average-sets of policy configuration. These policy "profiles" allow users to select a default set of policy values that line up with the average case, as presented by the application population. Policies can be promoted at an application level or at a group level. An XML approach is presented for representing the policy profiles. The approach allows for easy profile comparison and merging. A storage mechanism is also presented that describes how these policies should be made persistent in a distributed data storage system.

In Chapter 10, Md. Rafiqul Islam et al. In this paper we present an effective and efficient spam classification technique using clustering approach to categorize the features. In our clustering technique we use VAT (Visual Assessment and clustering Tendency) approach into our training model to categorize the extracted features and then pass the information into classification engine. We have used WEKA (www.cs.waikato.ac.nz/ml/weka/) interface to classify the data using different classification algorithms, including tree-based classifiers, nearest neighbor algorithms, statistical algorithms and AdaBoosts. Our empirical performance shows that we can achieve detection rate over 97%.

In Chapter 11, Tursun Abdurahmonov et al. This paper describes the use of Elliptic Curve Cryptography (ECC) over Prime Field (__) for encryption and digital signature of smart cards. The concepts of ECC over prime field (__) are described, followed by the experimental design of the smart card simulation. Finally, the results are compared against RSA algorithms.

In Chapter 12, Md. Golam Rabbani et al. The paper depicts a converged architecture of WiMAX and WiFi, and then proposes an adaptive resource distribution model for the access points. The resource distribution model ultimately allocates more time slots to those connections that need more instantaneous resources to meet QoS requirements. A dynamic splitting technique is also presented that divides the total transmission period into downlink and uplink transmission by taking the minimum data rate requirements of the connections into account. This ultimately improves the utilization of the available resources, and the QoS of the connections. Simulation results show that the proposed schemes significantly outperform the other existing resource sharing schemes, in terms of maintaining QoS of different traffic classes in an integrated WiMAX/WiFi architecture.

In Chapter 13, Yuchuan Wu et al. This paper presents a human daily activity classification approach based on the sensory data collected from a single tri-axial accelerometer worn on waist belt. The classification algorithm was realized to distinguish 6 different activities including standing, jumping, sitting-down, walking, running and falling through three major steps: wavelet transformation, Principle Component Analysis (PCA)-based dimensionality reduction and followed by implementing a radial basis function (RBF) kernel Support Vector Machine (SVM) classifier.

In Chapter 14, Subhasis Mukherjee et al. In this paper, we propose a reinforcement learning based approach that uses a dynamic state-action mapping using back propagation of reward and Q-learning along with spline fit (QLSF) for the final choice of high level functions in order to save the goal. The novelty of our approach is that the agent learns while playing and can take independent decision which overcomes the limitations of rule-base system due to fixed and limited pre-defined decision rules. The spline fit method used with the nose camera was also able to find out the location and the ball distance more accurately compare to the IR sensors. The noise source and near and far sensor dilemma problem with IR sensor was neutralized using the proposed spline fit method. Performance of the proposed method has been verified against the bench mark data set made with Upenn'03 code logic and a base line experiment with IR sensors. It was found that the efficiency of our QLSF approach in goalkeeping was better than the rule based approach in conjunction with the IR sensors. The QLSF develops a semi-supervised learning process over the rule-base system's input-output mapping process, given in the Upenn'03 code.

It is our sincere hope that this volume provides stimulation and inspiration, and that it will be used as a foundation for works yet to come.

July 2011

Morshed U. Chowdhury
Siddheswar Ray Monash

Contents

List of Contributors

Tursun Abdurahmonov
Multimedia University, Malaysia
tursun.abdurahmono07@mmu.edu.my

Iftekhar Ahmad
Edith Cowan University, Australia
i.ahmad@ecu.edu.au

Humayra Binte Ali
Flinders University
ali0041@flinders.edu.au

Ronghua Chen
Deakin University, Australia

Belal Chowdhury
La Trobe University, Australia

Morshed U. Chowdhury
Deakin University, Australia
morshed.chowdhury@deakin.edu.au

Loretta Davidson
Central Michigan University, USA
david11j@cmich.edu

Clare D'Souza
La Trobe University, Australia

Iqbal Gondal
Monash University, Australia
iqbal.gondal@monash.edu

Michael Hannaford
University of Newcastle, Australia

Md Rafiul Hassan
Department of Information and
Computer Science, KFUPM,
Dhahran, KSA
mrhassan@kfupm.edu.sa

Frans Henskens
University of Newcastle, Australia

Gongzhu Hu
Central Michigan University, USA
hu1g@cmich.edu

Shamsul Huda
University of Ballarat, Australia
s.huda@ballarat.edu.au

Helmi Mohamed Hussain
Multimedia University, Malaysia
helmi.hussain@mmu.edu.my

Naohiro Ishii
Aichi Institute of Technology, Japan
ishii@aitech.ac.jp

Md Rafiqul Islam
Deakin University, Australia
rislam@deakin.edu.au

Joarder Kamruzzaman
Monash University, Australia
joarder.kamruzzaman@monash.edu

Richard Leibbrandt
Flinders University, Australia

Trent Lewis
Flinders University, Australia

Tsuyoshi Miyazaki
Kanagawa Institute of Technology,
Japan
miyazaki@ic.kanagawa-it.ac.jp

Subhasis Mukherjee
University of Ballarat, Australia
smukherjee@academic.mit.edu.au

Toyoshiro Nakashima
Sugiyama Jogakuen University,
Japan
nakasima@sugiyama-u.ac.jp

David M W Powers
Flinders University, Australia

Md. Golam Rabbani
Monash University, Australia
golam.rabbani@monash.edu

Atul Sajjanhar
Deakin University, Australia
atul.sajjanhar@deakin.edu.au

Pinaki Sarkar
Jadavpur University, India
pinakisark@gmail.com

Mary F.H. She
Deakin University, Australia

Nasreen Sultana
Frankston Hospital, Australia

Mark Wallis
University of Newcastle, Australia
mark.wallis@uon.edu.au

Xianjing Wang
Deakin University, Australia
xianjing.wang@deakin.edu.au

Yuchuan Wu
Deakin University, Australia

John Yearwood
University of Ballarat, Australia
j.yearwood@ballarat.edu.au

Eng-Thiam Yeoh
Multimedia University, Malaysia
etyeoh@mmu.edu.my

Ming Zhang
Christopher Newport University,
USA
mzhang@cnu.edu

Zhongwei Zhang
University of Southern Queensland,
Australia
zhongwei@usq.edu.au

Hong Zhou
University of Southern Queensland,
Australia
hong.zhou@usq.edu.au

Analysis of ISSP Environment II Survey Data Using Variable Clustering

Loretta Davidson and Gongzhu Hu

Abstract. Social informatics, as a sub-field of the general field of informatics, deals with processing and analysis of data for social studies. One of the social data repositories is the International Social Survey Program (ISSP) the provides cross-national surveys on various social topic. Previous studies of this data often used a subset of available variables and sometimes a reduced number of records, and most of these analyses have focused on predictive techniques such as regression. In this paper, we analyze the Environment II module of this data set using variable clustering to produce meaningful clusters related to questionnaire sections and provide information to reduce the number of demographic variables considered in further analysis. Case level clustering was attempted, but did not produce adequate results.

Keywords: Social informatics, variable clustering, case level clustering.

1 Introduction

The International Social Survey Program (ISSP) [2], started in 1984, is a continuing annual program of cross-national (currently members from 43 countries) collaboration on surveys covering topics important for social science research. It contains annual modules that provide cross-national data sets on topics of social interests. Some of the most surveyed topics are listed in Table 1.

Loretta Davidson
Data Mining Program, Central Michigan University, Mt. Pleasant, MI 48859, USA
e-mail: david1lj@cmich.edu

Gongzhu Hu
Department of Computer Science, Central Michigan University,
Mt. Pleasant, MI 48859, USA
e-mail: hu1g@cmich.edu

R. Lee (Ed.): Software Eng., Artificial Intelligence, NPD 2011, SCI 368, pp. 1–13.

Table 1 ISSP Modules (1985 – 2012)

Topic	Years
Role of government	1985, 1990, 1996, 2006
Family, changing gender roles	1988, 1994, 2002, 2012
Religion	1991, 1998, 2008
Environment	1993, 2000, 2010

Environment was the subject of the 1993 and 2000 ISSP modules and is currently the topic for 2010 [1]. This paper is concerned with analyzing the Environment II (2000) data set using data mining techniques of variable clustering and case clustering. Previous analyses of the ISSP Environment 1993 and/or 2000 data have applied correlation, regression, factor analysis, and structural equation modeling techniques on a subset of the available variables [3, 4, 6, 7, 11]. In addition, these studies have focused on causal relationships. The analysis presented in this paper seeks to maximize the use of input variables by allowing variable clustering to segment variables and choose the most important variable from each cluster. Rather than taking a predictive modeling approach, this paper focuses on pattern recognition by implementing clustering of case level data.

2 Data Description

The ISSP Environment II data set, description, questionnaires, and codebook are available at the Interuniversity Consortium for Political and Social Research (ICPSR) at University of Michigan [8]. The raw data set consists of 31,402 cases and 209 variables. ISSP data is also accessible via GESIS Leibniz Institute for the Social Sciences that provides the variable overview, source questionnaire, and monitoring report documents. The Environment II questionnaire consists of 69 questions covering the following areas as stated in the overview and source questionnaire [5]:

- Left-right dimension
- Postmaterialism
- Attitudes towards science and nature
- Willingness to make trade-offs for environment
- Environmental efficacy
- Scientific and environmental knowledge
- Dangers of specific environmental problems
- Environmental protection, locus of control, effort
- Positive trade-off of environmentalism
- Trust information sources on causes of pollution
- Respondent's behaviors and actions to protect the environment
- Belief about God

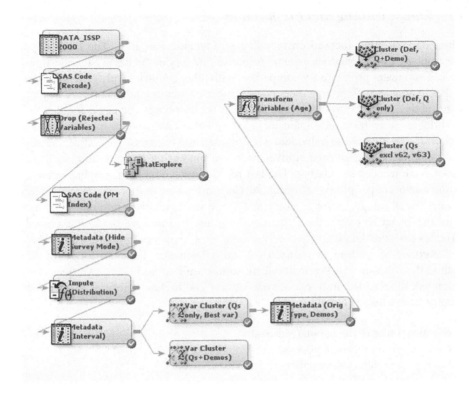

Fig. 1 SAS Enterprise Miner Diagram for ISSP Environment II data analysis.

- Type of area where respondent lives
- Grid-group theory

Most of the questions use a five-point or four-point ordinal scale although some are binary or nominal. Of the 209 variables, 118 were excluded from our analysis because they were country specific or for reasons related to usability, comparability, and redundancy. The remaining variables include 69 questionnaire variables, 19 demographic variables, a respondent ID, a weight variable, and a survey mode variable.

3 Methodology

We used the SAS Enterprise Miner software to run our analysis that consists of three parts: data preparation, variable clustering, and case clustering. The Enterprise Miner diagram developed for the analysis is shown in Fig. 1. In this diagram, each box is called a *node* representing a software module to perform a specific task. An arrow from node A to node B means the output produced by A is fed as an input to node B. We shall discuss the methods of each part in detail in this section.

3.1 Data Cleaning and Preparation

Initial data cleaning focused on recoding problematic variables. The knowledge questions for Chile contained an extra response category not in the master question- naire. This creates problems for comparison with other countries and the distribution of these variables differs from other countries. Since merging the incorrect category with another category would distort the data, the responses for Chile for this set of variables were removed. Spain included an extra category for three questions concerning the role of the individual, government, and business. The extra category was "neither" which seemed equivalent to not answering the question, so these values were recoded as missing. The last recode involved a demographic variable for the number of people in the household. Categories for 6 or more people varied by country, so all categories for 6 people and above were collapsed into one category. After the initial recoding, the Drop node was used to remove the 118 excluded variables from the data set.

A second SAS Code node was used to create a Postmaterialist index following the method described in [6]. Postmaterialism values are assessed using two questions which ask the respondent to choose the 1st and 2nd highest priorities for his/her country from a list of four choices:

1. Maintain order in the nation (Materialist)
2. Give people more say in government decisions (Postmaterialist)
3. Fight rising prices (Materialist)
4. Protect freedom of speech (Postmaterialist)

The combination of a person's rankings of the 1st and 2nd highest priorities indicates his/her materialist or postmaterialist preferences and is coded as an integer 1 – 4 shown in Table 2. The code, as the value of a new variable PM_INDEX, is interpreted as

1. Strong materialist
2. Weak materialist
3. Weak postmaterialist
4. Strong postmaterialist

It replaces the answers to the two original questions.

The next step of the data preparation uses a Metadata node to change variable roles and types. In this case, metadata was used to hide the survey mode variable which captures how the survey was administrated. Although this variable was not used in the analysis, we kept it in the data set for possible future use.

The Impute and second Metadata nodes were used to prepare the data for the Var Cluster node to perform variable clustering. The imputation method used for both interval and class (ordinal, nominal) data computes random percentiles of a variable's distribution to replace missing data. The Metadata node following the Impute node changes all of the variable types to interval. The Var Cluster node manages non-interval inputs by creating a new variable for each nominal or ordinal

Table 2 postmaterialist index

Highest Priority		Code
1st	2nd	
Materialist	Materialist	1
Materialist	Postmaterialist	2
Postmaterialist	Materialist	3
Postmaterialist	Postmaterialist	4

level. It was suggested changing variable types when many of the inputs are non-interval since the efficiency of the algorithm diminishes when there are over 100 variables [12].

3.2 Variable Clustering

Variable clustering is a method for segmenting variables into similar groups. It can be a useful tool for variable reduction. We uses two Var Cluster nodes each with its own objective. The Var Cluster node labeled "Qs only" is used to assess whether the questionnaire variables group according to topic. If so, the best variable from each variable cluster can be exported for use in case level clustering. The second Var Cluster node labeled "Qs+Demos" is used to analyze how the demographic variables behave when clustered with the questionnaire variables. If demographic variables pair with each other or questionnaire variables, perhaps some can be excluded from further analysis. Using variable clustering prior to case clustering will reduce the number of variables used in cluster analysis.

The variable clustering algorithm is a divisive hierarchical clustering technique that aims to maximize the explained variance [12]. At each stage the weakest cluster is selected for splitting until the stopping criteria is reached. Correlation was used for the clustering source matrix and variation proportion equal to 0.40 was the stopping criteria. The algorithm continues to split clusters until the variation explained by each cluster is at least equal to the stopping criteria.

The best variable from each cluster was exported from the Var Cluster "Qs only" node. The best variable has the lowest $1 - R^2$ ratio. In the following Metadata node, these variables are changed back to their original type and the demographic variables added to the analysis are changed to have an input role. At this stage, *Age* was the only continuous variable. The Transform Variables node changes it into a new nominal variable using 10 quantiles.

3.3 Clustering

The Ward method was used in the Cluster node. It follows the traditional hierarchical clustering algorithm that starts with n clusters for the n data records where each cluster containing a single data record. The algorithm repeatedly *merges* two clusters

Table 3 Clusters for Questionnaire Variables

Cluster	Description	Number of Variables.
1	Perceived threats	8
2	Environmental concern (unfavorable)	3
3	Trust in information sources except business/industry	5
4	Knowledge	3
5	Willingness to sacrifice	4
6	Actions (binary - group, money, petition, demonstration)	4
7	Attitudes - sacredness nature; God; science (unfavorable)	4
8	Locus of control and effort (Govt to others)	3
9	*Mixed* - country effort, trust bus/ind, private enterprise	3
10	Values/Grid-Group (egalitarian, anti-individualism)	3
11	Actions (ordinal - recycling, cut back on driving)	2
12	Attitudes - science and economic growth (favorable)	2
13	Perceived threat - nuclear power stations	2
14	*Mixed* - evolution T/F, population growth, intl agreements	3
15	Attitudes - modern life and economic growth (unfavorable)	3
16	Values/Grid-Group (egalitarian, communitarian)	2
17	Values/Grid-Group (hierarchy)	2
18	Demographic - describe where you live	1
19	Environmental Efficacy and Values/Grid-Group (fatalism)	4
20	Effort - business/industry, people, both	1
21	Effort - people, government, both	1
22	Knowledge - antibiotics kill bacteria not viruses	1
23	Values/Grid-Group - world is getting better	1
24	*Mixed* - animal testing and poor countries less effort	2
25	Values - Materialist-Postmaterialist index	1

selected based on an objective function until a specified measure is reached (such as a desired number of clusters, or a merging threshold). The objective function used in the Ward algorithm is to minimize the *total error sum of squares* when clusters C_a and C_b are selected to be merged, given in Equation (1).

$$\mathscr{T}(C_a, C_b) = ESS(C_{ab}) - ESS(C_a) - ESS(C_b) \tag{1}$$

where C_{ab} is the combined cluster after merging C_a and C_b, and $ESS(\cdot)$ is the *error sum of squares* of a given cluster, as defined in Equation (2) for the data set C.

$$ESS(C) = \sum_{i=1}^{n} ||\mathbf{x}_i - \mathbf{m}_C||^2 \tag{2}$$

where $n = |C|$, $\mathbf{x}_i \in C$, and $\mathbf{m}_C$ is the mean of the data records in C.

The **Cluster** node defaults apply rank encoding for ordinal data and GLM encoding for nominal data. Rank encoding uses the adjusted relative frequency for each level to create a new variable on a 0 to 1 scale. The nominal encoding technique create a new dummy variable for each level of the original variable. The initial cluster seed method was left as default which is the k-means algorithm [10].

Three **Cluster** nodes were run in the analysis: questionnaire variables and demographic variables, questionnaire variables only, questionnaire variables excluding two that relate to demographic information.

4 Results

4.1 Variable Clustering

The variable clustering for questionnaire and demographic variables included 86 input variables and produced 35 variable clusters. The total proportion of variation was 0.5582. Five clusters contained a single demographic variable which may indicate more importance or independence. Four clusters contained both demographic and questionnaire variables, while three clusters were all demographic variables. After reviewing the results, eight demographic variables were excluded from further analysis which left ten remaining demographic variables.

The variable clustering of questionnaire variables contained 68 input variables and produced 25 variable clusters. The total proportion of variation explained was 0.5333. Most clusters are centered on a questionnaire topic and variables had similar type. A description for each variable cluster is in Table 3. Three clusters contained variables of mixed types without a clear topic connection.

The make-up of mixed groups can be further investigated by using cross tabulation to find the intersection of variable levels with the highest frequency. For example, cluster 24 contained the attitude questions "It is right to use animals for medical testing if it might save human lives" and "Poorer countries should be expected to make less effort than richer countries to protect the environment" [5]. Both questions used a 5-point Agree/Disagree response scale. The cross tabulation reveals that 15% of respondents are in the "Agree (2)" category for both questions and that 13% "Agree (2)" to the animal testing, but "Disagree (4)" to less effort for poorer countries. Thus, this cluster can be described as agreement to animal testing with conflicted agreement for poor countries making less effort.

For each variable, the `Var Cluster` node output provides statistics for R^2 with own cluster, R^2 with next cluster, and $1 - R^2$ ratio. These statistics can be examined with plots to assess the strength or weakness of variables in a cluster. Fig. 2 shows R^2 with next cluster plotted against R^2 with own cluster. A high R^2 with own cluster and low R^2 with next cluster is desired.

Fig. 3 shows R^2 with next cluster plotted against $1 - R^2$ ratio. Lower $1 - R^2$ ratio indicates a better fit within the cluster.

The **Var Cluster** node results provide a cluster plot, shown in Fig. 4, which visualizes the relationship between clusters and variables according to relative

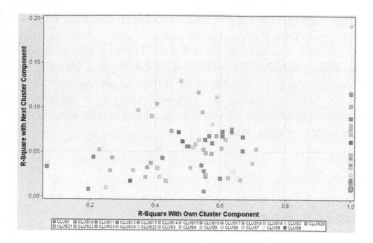

Fig. 2 R^2 with next vs. R^2 with own

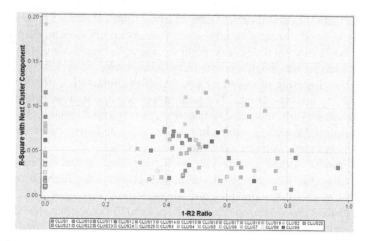

Fig. 3 R^2 with next vs. $1 - R^2$ ratio

distance, size, and orientation. Highlighted in the lower right side of Fig. 4 are clusters for willingness to sacrifice (cluster 5), actions (clusters 6 and 11), and postmaterialist values (cluster 25). This may support the research in [6] that identifies relationships between values, willingness to sacrifice, and actions.

4.2 Case Clustering

A total of 35 variables were considered in the case level cluster analysis. Questionnaire variables exported from the variable clustering accounted for 25 of the variables and 10 were supplementary demographic variables. Of these 35 variables, 18 were ordinal, 14 were nominal, and 3 were binary.

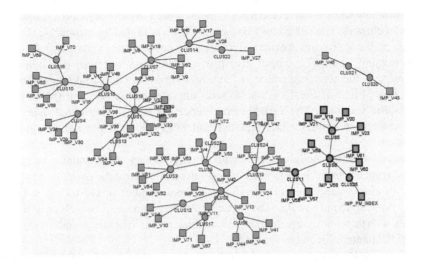

Fig. 4 Cluster Plot.

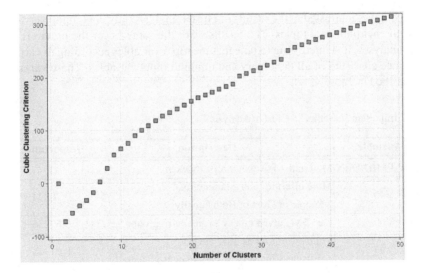

Fig. 5 Cubic Clustering Criterion Plot.

The first Cluster node with both demographic and questionnaire variables produced poor results. The algorithm terminated at the default maximum number of clusters instead of finding an optimal number of clusters. The cubic clustering criterion plot in Fig. 5 plots Cubic Clustering Criterion (CCC) against the number of clusters.

The plot does not show a local CCC peak which would indicate an optimal number of clusters. The cubic clustering criterion "is obtained by comparing the observed R^2 to the approximate expected R^2 using an approximate variance-stabilizing transformation" [12]. The output also provides information from decision tree modeling with the cluster segment as the target variable. The variable importance output showed that of the 13 variables with importance greater than 0.5, 8 were demographic variables. The variable importance measure is computed using the sum of squares error with the interpretation that values closer to one indicate greater importance.

A second Cluster node was run with supplementary demographic variables excluded from the analysis. Again, the algorithm terminated at the default maximum number of clusters and did not provide an optimal number of clusters. The CCC plot did not display any peaks. The variable importance output, which only had 3 variables with importance greater than .50, consisted of two binary variables and the nominal Postmaterialist index variable.

A third Cluster node was run which dropped two additional variables from the analysis. These two variables were part of the questionnaire but pertained to subjects that could be considered demographic information: belief in God and description of where the respondent lives. Although these variables did not reveal an important influence, I was interested in the results of cluster analysis only containing variables related to environmental topics. The results were the same as for the previous two cluster analyses. It is interesting to note that the top 5 variables according to variable importance consisted of all the binary and nominal input variables. These variables are described in Table 4.

Table 4 Important Variables in Cluster Analysis

Variable	Description	Importance
IMP_PM_INDEX	Materialist-Postmaterialist Index	1
IMP_V45	More effort to look after enviro: People or Govt or Both equally	0.70894
IMP_V60	Last 5 yrs given money to an enviro group: Yes / No	0.69236
IMP_V43	More effort to look after enviro: Business/Industry or People or Both equally	0.67948
IMP_V41	Enviro protection by Business/Insustry: Decide themselves or Govt pass laws	0.52280

A rerun of the last cluster node with user-specified 7 clusters produced clusters that can be described mainly in terms of the variables listed in Table 4; however, clusters defined by these inputs do not provide novel or insightful distinctions between groups.

5 Related Work

There are many research perspectives regarding social science survey data. Some researchers are interested in how values influence the formation of attitudes while others want to know if attitudes can predict behavior. The ISSP Environment II survey data allows for a wide range of analyses by including questions related to values, cultural theory, attitudes, knowledge, risk perception, willingness to sacrifice, and actions.

A study by [7] concluded that pro-environmental attitudes were poor predictors of pro-environmental behaviors based on regression analysis of 1993 ISSP Environment I data for New Zealand. This analysis was from a marketing perspective which reasoned that the effort to measure attitudes is fruitless unless a strong causal link between attitudes and behaviors is established.

In the study [3], six questions from the 1993 ISSP Environment I survey were included to compare values and pro-environmental actions in Western (Netherlands, United States) and Asian (Japan, Thailand, Philippines) countries. This study used factor analysis on Schwartz general values and progress/environment preferences (ISSP questions) and studied the relationship between them. The value and progress/environment factors were then used to predict three sets of pro-environmental behaviors using regression analysis. Differences in both value construction and predictors of behavior were found between Asian and Western countries.

A complex analysis by [11] used structural equation modeling to study the relationships between values (Schwartz values, Postmaterialism) and attitudes reflecting environmental concern, between concern, perceived threats, perceived behavior control and willingness to sacrifice, between willingness to sacrifice and pro-environmental behavior, and between values and behavior. This study used Schwartz harmony value data in addition to ISSP Environment II data for all available countries. The social science background motivating the analysis includes Schwartz values, Postmaterialism, and Value-Belief-Norm theory.

A study by [6] of the 1993 Environment I data for Norway compared the predictive ability of Postmaterialism and cultural theory on attitudes. An index was created from two questions to measure the degree of Postmaterialist preferences. Postmaterialism is a value theory developed by Inglehart that identifies a shift in individual values in countries with economic stability [9]. The values shift from being dependent on basic material security to quality of life issues and individual liberty [6, 3, 11]. The Norwegian questionnaire also included eight cultural theory questions that were grouped into four cultural bias dimensions. Cultural theory was developed by Douglas and is a typology that groups individuals according to "group" and "grid" dimensions. Group refers to the degree of membership in a social unit and grid refers to an individual's ability to negotiate relationships [13]. There are four dimensions of cultural bias depending on whether the group and grid factors are strong or weak. These grid-group dimensions are labeled hierarchy, egalitarianism, individualism, and fatalism [13]. The study by [6] performed factor analysis on twenty-nine attitude questions in the ISSP Environment I data and the resulting factors were used as dependent variables in regression analysis with the Postmaterialist index and cultural bias dimensions as independent variables.

6 Conclusion

While the variable clustering produced meaningful clusters, the case clustering did not. The questionnaire variables are mostly ordinal type which may not be optimal for case level clustering in this situation if there are not marked distinctions among the response levels. Since the variable clustering produced good results, perhaps the cluster components could be used in predictive analysis with action or attitude components as dependent variables. Cluster components were not used in the present analysis since their interpretation is not as straightforward, but using cluster components in future analyses would maximize the contribution from all variables. In addition, this study only focused on the second wave of the ISSP Environment module. The strength of the variable clustering technique for this type of data could be assessed by repeating the analysis with 1993 data. One would need to consider the differences between the 1993 and 2000 questionnaires in comparing the results.

Overall, the variable clustering supports a distinction among questionnaire topics. Given this information, analyses should be broadened to include input variables from more areas instead of focusing on a handful of selected variables. Perhaps this would lead to a more comprehensive understanding of research results from social science survey data.

References

1. Archive and data. International Social Survey Programme,
   http://www.issp.org/page.php?pageId=4
2. International Social Survey Programme (2010), http://www.issp.org
3. Aoyagi-Usui, M., Vinken, H., Kuribayashi, A.: Pro-environmental attitudes and behaviors: An international comparison. Human Ecology Review 10(1), 23–31 (2003)
4. Franzen, A.: Environmental attitudes in international comparison: An analysis of the ISSP surveys 1993 and 2000. Social Science Quarterly 84(2), 297–308 (2003)
5. ISSP Environment II data documentation (2009),
   http://www.gesis.org/en/services/data/survey-data/issp/
   modules-study-overview/environment/2000/
6. Grendstad, G., Selle, P.: Cultural theory, postmaterialism, and environmental attitudes. In: Ellis, R.J., Thompson, M. (eds.) Culture Matters: Essays in Honor of Aaron Wildavsky, pp. 151–168. Westview Press (1997)
7. Hini, D., Gendall, P., Kearns, Z.: The link between environmental attitudes and behaviour. Marketing Bulletin 6, 22–31 (1995)
8. ICPSR study No. 4104 International Social Survey Program: Environment II, 2000 (2009), http://www.icpsr.umich.edu

9. Inglehart, R.: The silent revolution in post-industrial societies. American Political Science Review 65, 991–1017 (1971)
10. Lloyd, S.P.: Least squares quantization in PCM. IEEE Transactions on Information Theory 28(2), 129–137 (1981)
11. Oreg, S., Katz-Gerro, T.: Predicting proenvironmental behavior cross-nationally: Values, the theory of planned behavior, and value-belief-norm theory. Environment and Behavior 38(4), 462–483 (2006)
12. SAS Institute Inc.: Enterprise Miner Help Documents (2009)
13. Thompson, M., Ellis, R., Wildavsky, A.: Cultural Theory, pp. 1–18. Westview Press (1990)

Polar Transformation System for Offline Handwritten Character Recognition

Xianjing Wang and Atul Sajjanhar

Abstract. Offline handwritten recognition is an important automated process in pattern recognition and computer vision field. This paper presents an approach of polar coordinate-based handwritten recognition system involving Support Vector Machines (SVM) classification methodology to achieve high recognition performance. We provide comparison and evaluation for zoning feature extraction methods applied in Polar system. The recognition results we proposed were trained and tested by using SVM with a set of 650 handwritten character images. All the input images are segmented (isolated) handwritten characters. Compared with Cartesian based handwritten recognition system, the recognition rate is more stable and improved up to 86.63%.

1 Introduction

Handwriting recognition is the capability of computer to translate or interpret human handwriting characters and display in different ways such as documents, graphics and other devices. Furthermore, it is an automated process in computer vision involving the pattern recognition and machine learning techniques for recognition of the human handwriting [1]. Typically, there are two different approaches in handwriting recognition: off-line and on-line. In on-line handwriting recognition, the machine recognizes the writing while the user writes on an electronic interface, such as an electronic pen or a mouse. The common on-line recognition method is that the pen-tip movements are picked up by a sensor based on x-y coordinates values. On the contrary, in off-line handwriting recognition, the optical character has usually been completed and converted as an image before performing handwriting recognition. In this paper, we merely concentrate on off-line handwriting recognition techniques. Handwriting character recognition (HCR) has received tremendous popularity in academic and commercial field. Despite enormous progresses in handwriting recognition, achieving high accuracy in the offline handwriting recognition is challenging nevertheless.

Xianjing Wang · Atul Sajjanhar
School of Information Technology, Deakin University
Burwood 3125, Victoria, Australia
e-mail: {xianjing.wang,atul.sajjanhar}@deakin.edu.au

R. Lee (Ed.): Software Eng., Artificial Intelligence, NPD 2011, SCI 368, pp. 15–24.
springerlink.com © Springer-Verlag Berlin Heidelberg 2011

The aim of this paper is to present a polar coordinate handwritten recognition system involving normalization, Polar transformation, feature extraction and SVM classification methodologies to achieve high recognition performance. The comparisons and evaluations for zoning feature extraction methods applied in Polar system are provided.

The remainder of this paper is organized as follows: Section 2 represents literature review in this field. Section 3 describes our proposed methodologies in the investigation and implementation. Section 4 shows the results for each set of experiments in this research and discusses and summarizes the procedures. Section 4 is the conclusion.

2 Background

A typical HCR system consists of five major stages: image pre-processing, segmentation, feature extraction, classification and post-processing.

2.1 Pre-processing

Pre-processing on the input image converts to a binary image, removes noise, segments lines, and normalization. In the character images pre-processing stage, following image pre-processing techniques are used in the experiment: 1) Binarization. 2) Noise reduction. 3) Skeletonization. 4) Resizing. 5) Zoning.

In general recognition system, the first step of processing usually consists of image enhancement and converting the grey level image to binary image as required in image pre-processing. Afterward converting image from the gray-scale image to binary format, the thresholding technique is used to separate the useful front pixels from the background pixels. A number of algorithms had been proposed in recent works in [2, 3, 4], for image enhancement and conversion of gray level image to binary image. Noise reduction is performed before or after binarization, which identifies and corrects the noise pixels. These sorts of techniques are based on image filtering theory and mathematical morphology [3]. Furthermore, a normalization step normalizes the handwriting sample images from varied stroke width. The methods generally apply to binary images and normalize the strokes width to single pixel thickness. Some approaches had been done in this field in [5].

2.2 Feature Extraction

Feature extraction plays an essential role in handwriting recognition. In HCR process, a text image must be either processed by feature extraction or feature processing after image pre-processing. The selected features will be classified for matching the best classifier [6].

There are many different feature extraction methods adequate for different representations of characters. Histogram features based on contour and skeletal representations of the image are commonly used in characters segmentation, feature representation and the rotated image [7]. An approach based on the horizontal projection histogram had been used for Arabic printed text in [8]. The drawbacks of projection histograms appear in the lose features of characters' shape. The shape of character does not exist with character after projecting histogram.

Zone based feature extraction method provides good result when certain preprocessing steps like filtering and smoothing are not considered. Different zoning mechanisms bring different recognition accuracy rates. In [9], the experiments were started with 4 equal zones, 5 horizontal zones, 5 vertical zones, and 7 zones. The confusion between characters was conducted in different number of zones. The experimental results were 82.89%, 81.75%, 80.94% and 84.73% for different number of zone mechanisms respectively. As we can notice from these experimental results, more number of zones does not mean better recognition rate. It is more based on the language type and writing style.

Rajashekararadhya and Ranjan [10] described an efficient zone based feature extraction algorithm on thinning binary characters. For extracting the features, each character image is divided into n equal zones. Average distance from the character image centroid to each pixel presents in each zone is computerized (Figure 3). The average distance between the zone centriod to each pixel presents in each zone is computed for the second features. Similarly, some zone based feature extraction methods were reported in [11, 12].

In view of zoning based feature extraction experiments, we investigated two zone based feature extraction algorithms inspired from the previous work in [10] in the polar transformation system.

2.3 Machine Classifier: Support Vector Machines

Support Vector Machines (SVM) is a discriminative classification method that modifies the data input by changing its representation through kernel functions and using a margin-based approach to do the classification [13][14]. The new representation is a projection of the input data onto a higher dimensional space, whereby a linear hyperplane is used to classify the data. It was introduced by Vapnik [13] and a very useful tool for non-linear classification. The transformation of data to higher dimensional space can be efficiently done using what is known as the kernel trick. The objectives of SVM are: (1) to maximize the margin M and (2) maximize the classification. In what follows, we assume that we are given a training set D of patterns where $x_i \in R^d$ is the input feature vector and

Table 1. Kernel function for developing SVM models

Kernel Function	Mathematical Formula
Linear	$K(x_i, x_j) = x_i{}^T x_j$
Polynomial	$K(x_i, x_j) = [(x_i \cdot x_j) + k]^d$, d is the degree of polynomial
Radial Basis Function (RBF)	$K(x_i, x_j) = exp\left(-\frac{\|x_i - x_j\|^2}{2\sigma^2}\right)$, σ is the width of RBF function

$y_i \in \{-1, +1\}$ is the target output class for $i = 1, ..., n$. The training data with output $+1$ are called *positive class* ($y > 0$), while the others are called *negative class* ($y < 0$). The discriminating function is to establish the equation: $D(x) = w \cdot x + b$ ($w, x \in R^d, b \in R$) where w is the weight vector defining the hyperplane and b is the contribution from the bias weights. With the classification constraints $y_i(w \cdot x_i + b) \geq 1$, all the points of the same class can be left on the same side while maximizing the distance between the two classes and the hyperplane [16]. Thus, support vector machine solves the following optimization problem:

$$\text{Minimize} \quad \frac{1}{2}\|w\|^2 + C\sum_{i=1}^{n}\varepsilon_i$$

$$\text{Subject to} \quad y_i(w \cdot x_i + b) \geq 1 - \varepsilon_i, \tag{1}$$

$$\varepsilon_i \geq 0, \qquad i=1,...,n.$$

where C is a constant parameter, called regularization parameter. It determines the trade-off between two terms. Maximizing the first term corresponds to minimizing the Vapnik-Chervonenkis dimension of the classifier and minimizing the second term controls the empirical risk [17].

Furthermore, $K(x_i, x_j) \equiv \emptyset(x_i)^T \emptyset(x_j)$ is called the kernel function. To construct SVMs, a kernel function must be selected. In this study, we investigated three kernels as shown in the Table 1.

3 Proposed Recognition System

The proposed methodologies consist of five major stages: pre-processing, polar transformation, normalization, feature extraction, and classification as shown in Figure 1. Segmentation stage is not included in paper by reason that all the images in training set are segmented handwritten characters and this research work merely focuses on image pre-processing, polar transformation, feature extraction and classifier techniques. The training set files contains two groups of different types of images. One is Cartesian coordinate based images downloaded from the 'mathworks.com' website in [18] and another one is Polar transformation images transformed from the Cartesian coordinate based images. Both of them contain 650 samples of hand drawn images of English characters with 25 images for each 26 alphabets.

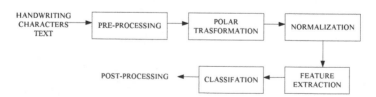

Fig. 1 Proposed Polar transformation system

3.1 Pre-processing

In the experiment, the input character images were converted to .bmp format referred to as monochrome or black and white from other standard format. After binarization, the noise pixels was identified and corrected for noise reduction by defining the threshold. The skeletonization defined the matrix to fit the entire character skeleton. Essentially, system resizes images into n equal sized zone which n is decided by the entire size of particular image. In place of the whole image, the selected feature extractions are applied onto individual zones.

3.2 Proposed Polar Transformation

We investigated the feature extraction and classification methods on Polar transformed images in the experiment. Polar mapping is an example of image warping which is similar to the standard Log Polar mapping had been used in computer vision research and in character recognition [19, 20]. The basic geometry of polar mapping is a conformal mapping from Cartesian coordinate points $f(x, y)$ in the image to points in the polar image $g(r, \theta)$. Two Cartesian coordinates x and y can be transformed to the polar coordinates radius (r) and angle (θ) by using the Pythagorean Theorem and Inverse Trigonometric Function. The mapping is described by:

$$r = \frac{\sqrt{y^2 + x^2}}{2}$$

$$\theta = arcsin\left(\frac{y}{r}\right), \ (x \geq 0) \tag{2}$$

Where x, y is the Cartesian coordinate of each point in character image. The r is defined as the radius of character image circle, and θ as the angle of the particular pixel in character image. A polar map aligns the Earth axis with the projection system, thus one of its poles lies at the map's conceptual center. We identify conceptual center and map an example image to polar image as shown in Figure 2. As input character images, all the character images in the training set are converted into Polar images before feature extraction and recognition processes.

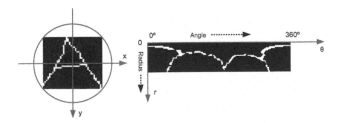

Fig. 2 Polar mapping example (Before Normalization)

3.3   Zoning Based Feature Extraction

In this paper, we focus on feature extraction methods extracted from binary images which can be obtained from a binarization of gray scale input image. We investigated two zoning based feature extraction methods inspired from [7, 10, 21].

We assume that the universe of discourse is selected, and image is divided
to 3 * 3 equal zones with equal size in advance, whereby extracting features from
individual zone rather than the whole image.

$$D = \frac{1}{N} \sum_{1}^{N} \sqrt{(x_c - x_i)^2 + (y_c - y_j)^2} \qquad (3)$$

Following expresses the procedures of two zone based feature extraction methods:
1) image centroid and zone based distance feature: computing average distance
from the character centroid to each pixel present in each zone (Figure. 3a). 2) zone
centroid and zone based distance feature: average distance from the zone centroid
to each pixel in each zone. As equation (3), $p(x_i, y_j)$ and $c(x_c, y_c)$ are the index-
es of each pixel and centroid respectively, and D presents the distance between the
particular pixel and centroid point (Figure 3b).

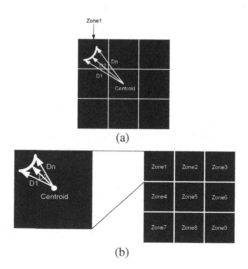

(a)

(b)

Fig. 3 (a) Image centroid and zone based distance, (b) Zone centroid and zone based
distance.

3.4   Classification

Scaling

Scaling before applying classification is very important [16]. A character image
divided into n equal zone would contain $n * 2$ uniform feature vectors used for

training and testing. The zone based features are expressed as float values in $[0, 1]$. In particular image or zone, if no pixel is involved in special zone, then the values is set as 0 in the feature vector. Image centroid based distance (ICD) and zone centroid base distance (ZCD) in a particular feature vector express as following (4) and (5) equation in regular case. n presents the number of zone divided in a image, and S^2 is the area of a image or a zone.

$$ICD = \frac{\sum_1^N \sqrt{(x_c - x_i)^2 + (y_c - y_j)^2}}{NS^2} \tag{4}$$

$$ZCD = \frac{n\sum_1^N \sqrt{(x_c - x_i)^2 + (y_c - y_j)^2}}{NS^2} \tag{5}$$

Support Vector Machines

In the experiment, we selected Polynomial kernel for non-linear modeling. We set $k = 1$ with degree 3 which maps a two dimensional input vector into a three dimensional feature space in a standard polynomial kernel form $K(x, y) = [(x \cdot y) + k]^d$. Furthermore, we used cross-validation to find the best values for the parameter C. The training set is first divide into n subsets of equal size in cross-validation. Then one subset is randomly selected to analysis and test on the remain of n-1 subset as training set. Each subset is analyzed once against the rest subsets to return a set of accuracy metrics. Classification experiment is performed on two training sets: Cartesian coordinate and Polar coordinate with 25*26 samples each.   10*26 samples are used for training and 15*26 for testing.

4   Experimental Result

The experiments were performed on two training sets: Cartesian coordinate images and Polar coordinate images. Both of them contain 650 English characters (A-Z) in total with 25 images for each 26 alphabets selected from database. All data are pre-processed in binary images. Fig. 4 and Fig. 5 show some Cartesian coordinate based and Polar transformed samples used for training and testing.

The table 2 and charts illustrate the recognition rate of each character and average recognition rate in this experiment. Table 2 presents zone-based feature extraction recognition results applying on Cartesian and Polar coordinate using Support Vector Machine classifier. In Figure 6 the confusion matrix of the SVM classfier is shown. Figure 7 demonstrate proposed recognition distribution rate (percentage: %) in Polar and Cartesian coordinates.

Fig. 4 Sample of handwritten characters

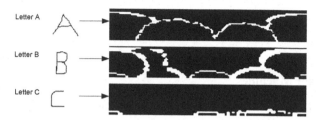

Fig. 5 Sample of Polar transformation (Before normalization)

To carry out the advantages and disadvantages of zoning feature extraction methodology in different coordinate system, we compared with different scheme from the experiment recognition results as shown in the table and figures. On the basis of experiment results, the proposed method obtains recognition rate at 86.63% by using SVM classifier.

Table 2. SVM Recognition Results

Kernel Function	Cartesian Coordination	Polar Coordination
Polynomial	80.62%	86.63%

In each character recognition rate, there are some confusions among characters such as 'A', 'R', and 'M' as a group, and 'D', 'O', and 'Q' are another confusion group (Figure 6). Zoning might not be suitable for general presentation in each digit zone such as different character represent same zone distance in particular zones. For instance, zone 1, 3, 5, 8 and 9 represent nearly the same average distance from centroid to each pixel in some training set samples such as letter 'A', 'M' and 'R'. Moreover, all of them appear empty in zone 8, which is another common feature in these character images. The common empty zone feature of 'A', 'M' and 'R' may lead a special zone feature of a character fail to be recognized and matched in the system. A similar situation is seen in letter 'O', 'D', and 'Q' appears resembling average distances in particular zones causes a rapidly decrease reaching to 48% in letter 'D' recognition rate. Except the decline gap in letter 'D' distribution, the general trend of ZB in Polar recognition has more stable effect compare with Cartesian coordinate highlighted between two red dash lines in Figure 7.

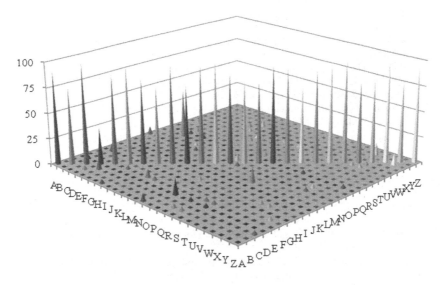

Fig. 6 Confusion matrix of SVM classifier in Polar system

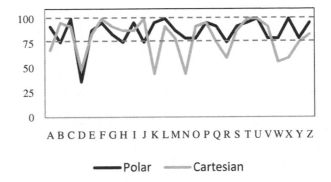

ABCDEFGHI JKLMNOPQRS TUVWXYZ

━━━Polar ━━━Cartesian

Fig. 7 Letter recognition distribution in Polar and Cartesian coordinate

The disadvantage of Polar transformation is that in some cases the transformed input images represent variety difference in skeleton as the different angle of each pixel. Especially, some rotated characters appear to increase differences due to the formal procedure which opens the image from the same coordinate point. It is clear to discover that the significant differences between Polar input image samples and original Cartesian image in Figure 5. Rotated image skeleton represents diversity in different zone region, which might be the main disadvantage for the Zoning-based distance calculation.

References

1. Liu, Z., Cai, J., Buse, R.: Handwriting recognition: soft computing and probabilistic approaches. Springer, Berlin (2003)
2. Deodhare, D., Suri, N.R., Amit, S.R.: Preprocessing and image enhancement algorithms for a form-based intelligent character recognition system. International Journal of Computer Science & Applications 2, 131–144 (2005)
3. Chandhuri, B.B.: Digital document processing: major direction and recent advances. Springer, London (2007)
4. Yan, C., Leedham, G.: Decompose-threshold approach to handwriting extraction in degraded historical document images, pp. 239–244. IEEE Computer Society, Los Alamitos (2004)
5. Nishida, H.: An approach to integration of off-line and on-line recognition of handwriting, pp. 1213–1219. Elsevier Science Inc., Amsterdam (1995)
6. Melin, P., Castillo, O., Ramirez, E.G., Janusz, K., Pedrycz, W.: Analysis and design of intelligent systems using soft conputing techniques. Springer, Berlin (2007)
7. Trier, D.O., Jain, K.A., Taxt, T.: Feature extraction methods for character recognition. Pattern Recognition 29, 641–662 (1996)
8. Pechwitz, M., Margner, V.: Baseline estimation for Arabic handwriting words. In: Proceedings Frontiers in Handwriting Recognition, pp. 479–484 (2002)
9. Simone, A.B.K., Cinthia, F.O.A.: Perceptual zoning for handwritten character recognition. In: Proceedings of 12th Conference of IGS, pp. 178–182 (2005)
10. Rajashekararadhya, S.V., Ranjan, P.V.: Efficient Zone Based Feature Extraction Algorithm for Handwritten Numeral Recognition of Rour Popular South Indian Script. Journal of Theoretical and Applied Information Technology, 1171–1181 (2005)
11. Majumdar, A., Chaudhuri, B.B.: Printed and handwritten bangla numeral recognition using multiple classifier outputs. In: Proceedings of the First IEEE ICSIP 2006, vol. 1, pp. 190–195 (2006)
12. Hanmandlu, M., Grover, J., Madasu, V.K.: Inout fuzzy for the recognition of handwritten Hindi numerals. In: International Conference on Informational Technology, pp. 208–213 (2007)
13. Vapnik, V.N.: The nature of statistical learning theory. Wiley, New York (1995)
14. Muller, K.R., Mika, S., Ratsh, G., Tsuda, K., Scholkopf, B.: An introduction to kernel-based learning algorithms. IEEE Trans. Neural Netw. 12(2), 181–201 (2001)
15. Camastra, F.: A SVM-based cursive character recognizer. Pattern Recognition Society 40, 3721–3727 (2007)
16. Hsu, C.W., Chang, C.C., Lin, C.J.: A Practical Guide to Support Vector Classification. Department of Computer Science. National Taiwan University (2003)
17. Kim, H., Pang, S., Je, H., Kim, D., Bang, S.Y.: Constructing support vector machine ensemble. Pattern Recognition 36, 2757–2767 (2003)
18. Dinesh, G.D.: Matlabcentral (December 2009),
    http://www.mathworks.com/matlabcentral/fileexchange/
    26158-hand-drawn-sample-images-of-english-characters
19. Nopsuwan chai, R.: Discriminative training method and their applications to handwriting recognition. Technical Report 652, University of Cambridge (2005)
20. Mpiperis, I., Malassiotis, S., Strintzis, M.G.: 3-D face recognition with the geodesic polar representation. IEEE Transaction on Information Forensics and Security, 537–547 (2007)
21. Blumenstein, M., Verma, B., Basli, H.: A Novel Feature Extraction Technique for the Recognition of Segmented Handwritten Characters. In: The 7th International Conference on Document Analysis and Recognition (ICDAR 2003), pp. 137–141 (2003)

Examining QoS Guarantees for Real-Time CBR Services in Broadband Wireless Access Networks

Hong Zhou and Zhongwei Zhang

Abstract. A wide range of emerging real-time services (e.g. VoIP, video conferencing, video-on-demand) require different levels of Quality of Services (QoS) guarantees over wireless networks. Scheduling algorithms play a key role in meeting these QoS requirements. A distinction of QoS guarantees is made between deterministic and statistical guarantees. Most of research in this area have been focused on deterministic delay bounds and the statistical bounds of differentiated real-time services are not well known. This paper provides the mathematical analysis of the statistical delay bounds of different levels of Constant Bit Rate (CBR) traffic under First Come First Served with static priority (P-FCFS) scheduling. The mathematical results are supported by the simulation studies. The statistical delay bounds are also compared with the deterministic delay bounds of several popular rate-based scheduling algorithms. It is observed that the deterministic bounds of the scheduling algorithms are much larger than the statistical bounds and are overly conservative in the design and analysis of efficient QoS support in wireless access systems.

1 Introduction

In recent years, there have been increasing demands for delivering a wide range of real-time multimedia applications (e.g. VoIP, video conferencing, video-on-demand) in broadband wireless access networks. IEEE 802.16 standard for broadband wireless access systems [18] provide fixed-wireless access

Hong Zhou
University of Southern Queensland, Toowoomba QLD 4350
e-mail: hong.zhou@usq.edu.au

Zhongwei Zhang
University of Southern Queensland, Toowoomba QLD 4350 Australia
e-mail: zhongwei@usq.edu.au

R. Lee (Ed.): Software Eng., Artificial Intelligence, NPD 2011, SCI 368, pp. 25–40.
springerlink.com   © Springer-Verlag Berlin Heidelberg 2011

for individual homes and business offices through the base station instead of cable and DSL in wired networks. This creates great flexibility and convenience as well as challenges for the design and analysis of such networks. Multimedia communications require certain level of Quality of Services (QoS) guarantees and individual applications also have very diverse QoS requirements. It requires the wireless access networks to support real-time multimedia applications with different QoS guarantees.

QoS performance is characterized by a set of parameters in any packet-switched network, namely end-to-end delay, delay variation (i.e. jitter) and packet loss rate [1, 2, 3]. Unlike non-real-time services, quality of real-time services is mainly reflected by their delay behaviors, namely, delay and delay variation. A distinction in QoS performance guarantees is made between *deterministic guarantees* and *statistical guarantees* [15]. In the deterministic case, guarantees provide a bound on the performance of all packets from a session. In other words, deterministic delay guarantees promise that no packet would be delayed more than D time units on an end-to-end basis. The value D is defined as the *Deterministic Delay Bound* (DDB). On the other hand, statistical guarantees promise that no more than a specified fraction, α, of packets would experience delay more than $D(\alpha)$ time units. $D(\alpha)$ is defined as the *Statistical Delay Bound* (SDB). As the fraction α becomes smaller, the statistical delay bound increases. In the special case of $\alpha = 0$, the statistical delay bound reaches the maximum value and is equal to the deterministic delay bound. That is, $D(\alpha) = D$.

Similarly, delay variation is defined as the difference between the best and worst case expectation of variable delays (i.e. mainly queueing delays). In statistical case, the best case is equal to zero and the worst case is a value likely to be exceeded with a probability less than α(for example 10^{-9}). It should be noted that, when the occasional exceptions are rare enough (e.g. $\alpha = 10^{-9}$), though the SDB may still be much smaller than DDB, the distinction between statistical guarantees and deterministic guarantees is negligible for most real-time services. Consequently, the QoS offered by statistical guarantees will be as good as those offered by deterministic guarantees for these real-time services. In general, the deterministic delay bound is much larger than the statistical delay bound and thus it is overly conservative. A statistical delay bound is sufficient for almost all real-time services.

Scheduling algorithms play a key role in satisfying these QoS requirements. In the past twenty years, a significant volume of research has been published in literature on scheduling algorithms such as Packet-by-packet Generalized Processor Sharing (PGPS) [4], Self-Clocked Fair Queueing (SCFQ) [5], Latency-Rate (LR) Server [6], Start-time Fair Queueing (SFQ) [7], Wireless Packet Scheduling (WPS) [8] and Energy Efficient Weighted Fair Queueing (E^2 WFQ) [9]. However, these research were basically focused on the deterministic delay bounds. The statistical delay bounds of scheduling algorithms meeting different QoS requirements have not been adequately studied.

In this paper, we examine the access delay of CBR real-time traffic in wireless access systems (e.g. IEEE 802.16). Future backbone networks have very high bandwidth and the delay experienced is very low. On the other hand, the access networks have relatively limited speed and the delay experienced by CBR traffic is very large. Therefore the analysis and design of the wireless access systems to support the QoS of real-time services is very important.

IEEE 802.16 standard for Broadband wireless access systems are designed to support a wide range of applications (data, video and audio) with different QoS requirements. IEEE 802.16 defines four types of service flows, namely Unsolicited Grant Service (UGS), real-time Polling Service (rtPS), non-real-time Polling Service (nrtPS) and Best effort service (BE). There are two types of service flows for real-time services, i.e. UGS supports CBR traffic including VoIP streams while rtPS supports real-time VBR flows such as MPEG video [13, 14]. IEEE 802.16 standard left the scheduling algorithm for the uplink and downlink scheduling algorithm undefined. Wongthavarawat and Ganz in [17] proposed a combination of strict priority scheduling, Earliest Deadline First [19] and WFQ [4]. The CBR real-time traffic, (UGS) has preemptive priority over other type of flows. In this paper, we are concerned with real-time CBR traffic and we assume there are different levels of QoS requirements within CBR traffic. For example, the emergence and remote medical CBR services should have higher QoS requirements than normal VoIP chats. We analyse the delay for different service levels of CBR traffic by solving class-based nD/D/1 queue.

The superposition of independent streams with periodic arrival patterns has been modelled by nD/D/1 queue in several past works [10, 11, 12, 16]. The problem of traffic delay analysis can be solved by finding the waiting time distribution of nD/D/1 Queue. Our study is different from the cited research as we differentiate CBR streams by priorities and analyse the delay for nD/D/1 queue with arbitrary number of service priorities in IEEE 802.16 broadband access networks.

The rest of this paper is organised as follows. In Section 2, a discrete-time P-FCFS queueing system model with Constant Bit Rate (CBR) inputs is defined and illustrated. In Section 3, we analyse the model in general cases that there are arbitrary number of priority levels and there are arbitrary number of traffic sources at individual levels. The queueing delay distribution for each service level is derived. In Section 4, we provide the delay distributions of different priority classes obtained by both mathematical and simulation analysis. Section 5 concludes the paper.

2   Discrete-Time Priority Queueing Model

The nD/D/1 model with several priority levels analyzed here has the following characteristics: (a) independent periodic sources with same period; (b) deterministic service/transmission time; (c) with priority levels; (d) discrete-time queueing system, or say slotted server.

As illustrated in Figure 1, we assume that there are totally N active real-time sources which are classified into K priority levels. For each priority level x $(1 \leq x \leq K)$, the number of sources is N_x. Each source generates fix length cells periodically with same period T. To keep the system stable, period T has to be greater than the total number of sources N.

The discrete-time model assumes slotted transmission on the path. The time axis is divided into fixed length slots and the transmission is restricted to start at a slot boundary. As a result of this restriction, each packet has to wait at least until the start of the next time slot.

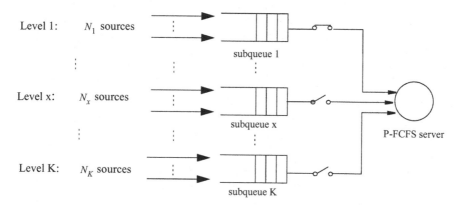

Fig. 1 Priority queueing model

Without loss of generality, time slot and cell length are assumed to be unity. Then the service time of each packet is also equal to unit time. In a P-FCFS queue, packets with higher priorities will be served first and those with lower priority will wait until all the higher priority packets have been served. The packets with the same priority will be served in FCFS principles. Thus, the delay experienced by a packet is equal to the number of packets found in the same and higher priority queues at the arrival time and packets with higher priority that have arrived between the arrival time and transmission time plus the remaining transmission time of the packet in service. Note that the packets from the lower priorities do not affect the delay experienced by a packet from higher priorities.

3 Mathematical Analysis

Let the source tested, say i, be the tagged source and all other sources be background sources. Suppose that the priority of the tagged source is x ($x =$

$1, 2, \cdots, K$). Because the sources with lower priorities than the tagged source do not affect the delay of the tagged source, we only consider the sources with the same or higher priorities than the tagged one. Let N_H be the total number of sources with higher priorities than the tagged one. Thus,

$$N_H = N_1 + \cdots + N_{x-1} \tag{1}$$

Let the waiting time/queueing delay experienced by the packet from the tagged source q_i be the interval from the beginning of the first slot since the packet arrives to that of the slot at which it starts to be served. Note that the residual slot period until the start of the next time slot is omitted and the delay is always an integer. In general this simplification does not effect our results. In what follows, we calculate the probability when the queueing delay q_i is equal to d, namely, $Pr\{q_i = d\}(d \geq 0)$.

Consider a period of time T from $t + d - T$ to $t + d$ and separate this interval into three sub-intervals. Suppose the arrival time of the tagged source i is uniformly distributed within the tth time slot ($[t - 1, t]$). The arrivals of background sources are independent and uniformly distributed in the interval $[t + d - T, t + d]$. The numbers of sources arriving on the sub-intervals are defined as follows (See Figure 2).

- n_H is the number of sources with higher priorities than arriving during $(t, t + d]$;
- n_x is the number of sources with priority x arriving during $(t, t + d]$;
- n_H' is the number of sources with higher priorities arriving during $(t - 1, t]$;
- n_x' is is the number of sources with the same priority arriving during $(t - 1, t]$;
- A_τ is the number of background sources with higher priorities arriving during the $t + \tau$th time slot;
- n_H'' is the number of sources with higher priorities arriving during $(t + d - T, t - 1]$, $n_H'' = N_H - n_H - n_H'$
- n_x'' is the number of sources with the same priority arriving during $(t + d - T, t - 1]$, $n_x'' = N_x - n_x - n_x'$.

Q_t is defined as the total length of packets waiting in higher priority sub-queues and packets in sub-queue x ahead of tagged packet at the end of tth time slot. When the queueing delay q_i is equal to d, the server will keep busy till $t + d$. Thus, the following two necessary and sufficient conditions must be satisfied.

$$Q_{t+d} = \max(Q_{t+d-1} - 1, 0) + A_d = 0 \tag{2}$$

$$\text{and} \quad Q_{t+\tau} > 0 \text{ for all } \tau = 0, 1, \cdots, d - 1 \tag{3}$$

From [3],

$$Q_{t+\tau} = max(Q_{t+\tau-1} - 1, 0) + A_\tau$$
$$= Q_{t+\tau-1} - 1 + A_\tau \tag{4}$$

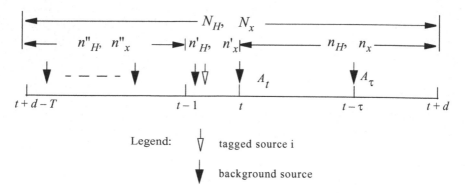

Legend: tagged source i

          background source

Fig. 2 The numbers of sources arriving in a period of time

By iteration on τ, $Q_{t+\tau}$ is equal to

$$Q_{t+\tau} = Q_t + A_1 + A_2 + \cdots + A_{\tau-1} + A_\tau - \tau$$
$$= Q_t + A(t, t + \tau) - \tau \qquad (5)$$

where $A(t, t+\tau)$ represents the total number of arrivals with higher priorities in the interval $[t, t + \tau]$. Then (2) and (3) are equivalent to

$$Q_t = d - n_H \qquad (6)$$
$$\text{and} \quad A(t, t + \tau) > \tau - Q_t \text{ for all } \tau = 0, 1, \cdots, d - 1 \qquad (7)$$

respectively. Thus, the probability of queueing delay is

$$Pr\{q_i = d\} = Pr\{Q_{t+\tau} > 0, \tau = 0, 1, \cdots, d - 1, Q_{t+d} = 0\}$$
$$= Pr\{A(t, t + \tau) > \tau - Q_t, Q_t = d - n_H\}$$
$$= \sum_{n_H=0}^{d-1} \sum_{n_x=0}^{N_x-1} \sum_{n'_H=0}^{N_H-n_H} Pr\{A(t, t + \tau) > \tau - Q_t, Q_t = d - n_H| \qquad (8)$$

$$A_H(t, t + d) = n_H, A_x(t, t + d) = n_x, A_H(t - 1, t) = n'_H\}$$
$$\cdot Pr\{A_H(t, t + d) = n_H, A_x(t, t + d) = n_x, A_H(t, t + d) = n'_H\}.$$

For the convenience of expression, in the following discussion, the term $Pr\{A(t, t + \tau) > \tau - Q_t, Q_t = d - n_H\}$ is used to represent $Pr\{A(t, t + \tau) > \tau - Q_t, Q_t = d - n_H|A_H(t, t + d) = n_H, A_x(t, t + d) = n_x, A_H(t - 1, t) = n'_H\}$ Moreover, $A_H(t, t + d)$, $A_x(t, t + d)$, and $A_x(t - 1, t)$ are independent random variables and (8) can be written

$$Pr\{q_i = d\} = \sum_{n_H=0}^{d-1} \sum_{n_x=0}^{N_x-1} \sum_{n'_H=0}^{N_H-n_H} Pr\{A(t, t+\tau) > \tau - Q_t, Q_t = d - n_H\}$$

$$\cdot Pr\{A_H(t, t+d) = n_H\} Pr\{A_x(t, t+d) = n_x\} Pr\{A_H(t-1, t) = n'_H\}.$$

$$= \sum_{n_H=0}^{d-1} \sum_{n_x=0}^{N_x-1} \sum_{n'_H=0}^{N_H-n_H} Pr\{A(t, t+\tau) > \tau - Q_t | Q_t = d - n_H\} \cdot Pr\{Q_t = d - n_H\}$$

$$\cdot Pr\{A_H(t, t+d) = n_H\} \cdot Pr\{A_x(t, t+d) = n_x\} \cdot Pr\{A_H(t-1, t) = n'_H\} \quad (9)$$

Let $\tau = d - \tau'$. Then,

$$A_H(t, t+d) = n_H - \sum_{j=1}^{\tau'} A_{d-j} > d - \tau' - d + n_H$$

which leads to

$$\sum_{j=1}^{\tau'} A_{d-j} < \tau' \quad \tau' = 1, 2, \cdots, d. \quad (10)$$

By the Ballot Theorem 4 in [12],

$$Pr\{\sum_{j=1}^{\tau'} A_{d-j} < \tau', \tau' = 1, 2, \cdots, d\} = 1 - \frac{n_H}{d}. \quad (11)$$

Substituting (11) int (9),

$$Pr\{q_i = d\} = \sum_{n_H=0}^{d-1} \sum_{n_x=0}^{N_x-1} \sum_{n'_H=0}^{N_H-n_H} Pr\{Q_t = d - n_H\} Pr\{A_H(t, t+d) = n_H\} (12)$$

$$\cdot Pr\{A_x(t, t+d) = n_x\} Pr\{A_H(t-1, t) = n'_H\}.$$

The next step is the evaluation of each term in (12) separately. The last three terms in (12) can be easily obtained from the *Probability Mass Function*(pmf) of Binomial random variables [17]. That is,

$$Pr\{A_H(t, t+d) = n_H\} = \binom{N_H}{n_H} \left(\frac{d}{T}\right)^{n_H} \left(1 - \frac{d}{T}\right)^{N_H-n_H} \quad (13)$$

$$Pr\{A_x(t, t+d) = n_x\} = \binom{N_x-1}{n_x} \left(\frac{d}{T}\right)^{n_x} \left(1 - \frac{d}{T}\right)^{N_x-n_x-1} \quad (14)$$

$$Pr\{A_H(t-1, t) = n'_H\} = \binom{N_H-n_H}{n'_H} \left(\frac{1}{T-d}\right)^{n'_H} \left(1 - \frac{1}{T-d}\right)^{N_H-n'_H-n_H} \quad (15)$$

In the following, the second term in (12) is calculated. Firstly, define Y as the rank of the tagged source i within the n_x' sources. Then

$$Q_t = \max(Q_{t-1} - 1, 0) + Y - 1 + n_H'. \tag{16}$$

For convenience of expression, let $L = max(Q_{t-1} - 1, 0)$. Thus,

$$Pr\{Q_t = d - n_H\} = Pr\{L + Y + n_H' - 1 = d - n_H\}$$

$$= \sum_{n_x'=1}^{N_x - n_x} Pr\{Y = d - n_H - n_H' - L + 1 | A_x(t-1,t) = n_x'\} Pr\{A_x(t-1,t) = n_x'\}$$

$$= \sum_{n_x'} \sum_{l} Pr\{Y = d - n_H - n_H' - L + 1 | L = l, A_x(t-1,t) = n_x'\} Pr\{L = l\}$$

$$\cdot Pr\{A_x(t-1,t) = n_x'\}. \tag{17}$$

Making use of the result shown in Appendix E of [10], the probability of the rank $Y = y$ among n_x' sources is simply the reciprocal of n_x'. That is

$$Pr\{Y = d - n_H - n_H' - L + 1 | L = l, A_x(t-1,t) = n_x'\} = \frac{1}{n_x'} \tag{18}$$

Thus

$$Pr\{Q_t = d - n_H\} = \sum_{n_x'} \sum_{l} \frac{1}{n_x'} Pr\{L = l\} Pr\{A_x(t-1,t) = n_x'\}. \tag{19}$$

Note that $L = d - n_H - n_H' - Y + 1$ and $1 \leq Y \leq n_x'$. If set $m = d - n_H - n_H'$, then,

$$max(m - n_x' + 1, 0) \leq l \leq m.$$

Thus,

$$m - n_x' + 1 \leq l \leq m \qquad \text{when } 1 \leq n_x' < m + 1$$
$$0 \leq l \leq m \quad \text{when } m + 1 \leq n_x' \leq N_x - n_x. \tag{20}$$

Equation (19) becomes

$$Pr\{Q_t = d - n_H\} = \sum_{n_x'=1}^{m} Pr\{A_x(t-1,t) = n_x'\} \sum_{l=m-n_x'+1}^{m} Pr\{L = l\}$$

$$+ \sum_{j_x=m+1}^{N_x-n_x} \frac{1}{n_x'} Pr\{L = l\} Pr\{A_x(t-1,t) = n_x'\}. \tag{21}$$

Noting that

$$\sum_{l=a}^{b} Pr\{L = l\} = Pr\{L > a - 1\} - Pr\{L > b\} \tag{22}$$

Applying (22) in (21) and combining the same terms, (21) becomes

$$Pr\{Q_t = d - n_H\} = \sum_{n'_x=1}^{m} \frac{1}{n'_x} Pr\{A_x(t-1,t) = n'_x\} Pr\{L > m - n'_x\}$$

$$+ \sum_{j_x=m+1}^{N_x-n_x} \frac{1}{n'_x} Pr\{A_x(t-1,t) = n'_x\} - \sum_{n_x=1}^{N_x-n_x} \frac{1}{n'_x} Pr\{A_x(t-1,t) = n'_x\} Pr\{L > m\}. \tag{23}$$

As shown in Appendix, the following can be obtained.

$$Pr\{Q_t = d - n_H\} = \frac{T-d}{N_x - n_x} (T-d)^{-(N_x-n_x)} (T-d-1)^{-n_H^{''}}$$

$$\left\{ \sum_{n'_x=1}^{N_x-n_x} \binom{N_x - n_x}{n_x} \#\Omega_{\leq m} [T-d-1, n''_{x+H}] - \sum_{n'_x=1}^{m} \binom{N_x - n_x}{n_x} \#\Omega_{\leq m-n'} [T-d-1, n''_{x+H}] \right\} \tag{24}$$

where $m = d - n_H - n'_H$, $n''_{x+H} = n''_x + n''_H = N_H + N_x - n_H - n'_H - n_x - n'_x$, and

$$\#\Omega_{\leq c}[a, b] = (a - b + c + 1) \sum_{j=0}^{c} \binom{b}{j} (-1 - c + j)^j (a + c - j + 1)^{b-j-1}. \tag{25}$$

Substituting (13), (14), (15) and (24) into (12), and canceling the like terms gives

$$Pr\{q_i = d\} = T^{-(N_H+N_x)+1} [U(d, N_H, N_x) - V(d, N_H, N_x)] \tag{26}$$

where $U(d, N_H, N_x)$ and $V(d, N_H, N_x)$ represent

$$U(d, N_H, N_x) = \sum_{n_H} \sum_{n_x} \sum_{n'_H} \Psi \sum_{n'_x=1}^{N_x-n_x} \binom{N_x - n_x}{n'_x} \#\Omega_{\leq m}[T-d-1, n''_{x+H}]$$

$$V(d, N_H, N_x) = \sum_{n_H} \sum_{n_x} \sum_{n_H} \Psi \sum_{n'_x=1}^{m} \binom{N_x - n_x}{n'_x} \#\Omega_{\leq m-n'_{n_x}}[T-d-1, n''_{x+H}]$$

and

$$\Psi = \frac{d^{n_H+n_x-1}}{N_x - n_x} (d - n_H) \binom{N_H}{n_H + n'_H} \binom{N_x - 1}{n_x} \binom{n_H + n_{H'}}{n_H}$$

In addition, it is known that

$$Pr\{q_i > d\} = 1 - \sum_{j=0}^{d} Pr\{q_j = n_x\} \tag{27}$$

Substituting (26) into (27), the tail distribution of queueing delay will be obtained.

4 Numerical Results

4.1 Mathematical Results

Based on the analysis in Section 3, logarithmic functions of the tail distributions of the queueing system at a load of 0.8, $LogPr\{q_i > d\}$, are calculated and plotted in Figure 3 using Mathematica. In Figure 3, the period is equal to $T = 50$ and there are 41 background sources (i.e. $N = 41$). The sources are classified into four priorities and each priority has an equal number of sources (excluding the tagged source) in the queueing model, namely, $N_x' = (N - 1)/4, x \in [1, 4]$. From Figure 3, it can be seen that delays are differentiated by their priorities and when the priority becomes lower, the delays become significantly larger.

For comparison, the tail distributions for the queueing delay of a queue with distinct priorities and a queue without priority are also plotted in Figure 4 and Figure 5 [15]. Figure 4 shows the delay distributions of sources

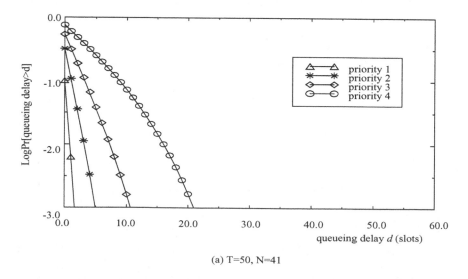

(a) T=50, N=41

Fig. 3 Delay distributions of nD/D/1 queue with four service classes(T=50,N=41)

of several priority levels when each source has a distinct priority level. The delay distribution of sources with similar priorities are very close to each other. For example, the delay difference between priority 9, 10 and 11, are negligible. Comparing Figure 6 with Figure 4, the gap between any different priority level is obvious. Thus, it is not necessary for every single source to have a distinct priority and three or four classes is enough for providing differentiated services.

The delay distributions of a queueing system without any priority is given in Figure 5. Comparing the curves in Figure 3 with the curve of $T = 50$ and

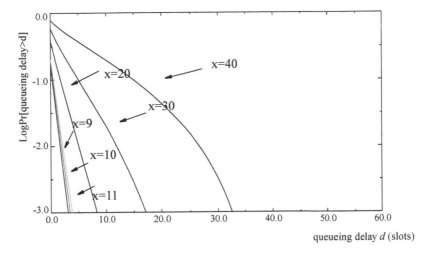

Fig. 4 Delay distributions of nD/D/1 queue with distinct priorities(T=50,N=41)

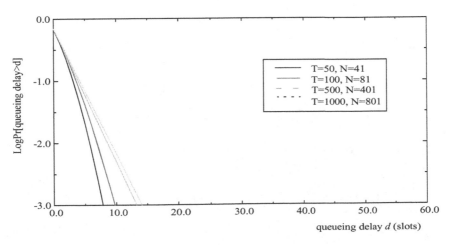

Fig. 5 Delay distributions of nD/D/1 queue without differentiated services

$N = 41$ case in Figure 5, the first two higher priorities have better performance than the one without priority. The third priority has a slightly worse performance than the one without priority. The fourth priority has obvious worse performance than the one without priority. Thus, priority queueing can provide individual sources with a desirable QoS according to their requirements. The network resources are efficiently and scientifically used.

4.2   Simulation Results

For demonstrating the correctness of our mathematical analysis, the queueing system discussed in the previous sections was implemented and analyzed in

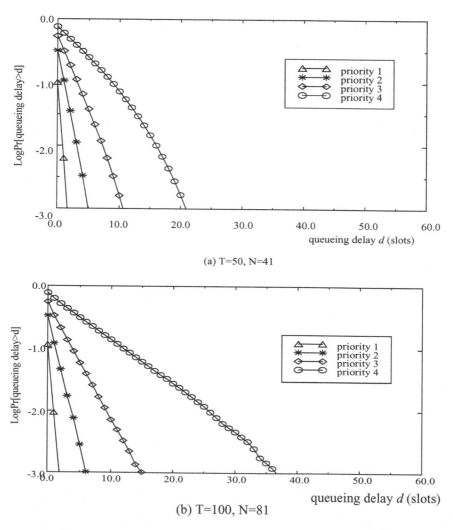

(a) T=50, N=41

(b) T=100, N=81

Fig. 6 Delay distributions of nD/D/1 queue with four service classes

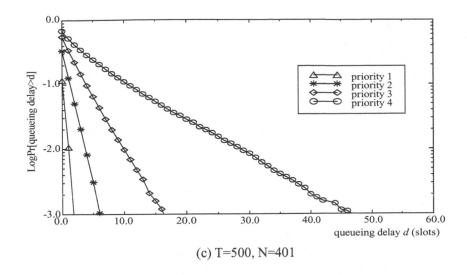

(c) T=500, N=401

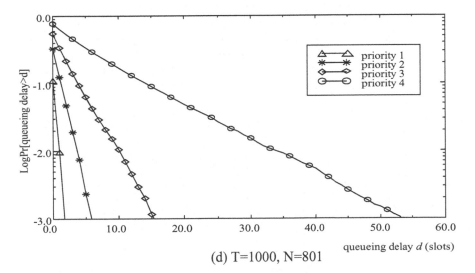

(d) T=1000, N=801

Fig. 6 (*continued*)

simulations using OPNET (Optimized Network Engineering Tool) package. By running the simulation model established in the studies, tail distributions for different numbers of sources were obtained and shown in Figure 6. The parameters of the queueing system in the simulation were set in the same manner to those used in the previous calculations.

Comparing of Figure 6 with Figure 3 shows that the delay distributions from mathematical analysis and simulation are identical.

Comparisons of Figure 6(a), (b), (c) and (d) show that as the number of sources increases, the delay for each class increases. This will be more obvious as the value of the vertical axis decreases. However, the increases in delay become slower as the number of sources increases, especially for higher priorities. Thus, the delay is statistically bounded even in the core network with a large number of competing sources.

4.3 Comparison with Deterministic Delay Bounds

This section compares the statistical delay bounds with the deterministic delay bounds (i.e. queueing latencies) of several rate-based scheduling algorithms. In this example, we use the same scenario as defined in the first example of Section 4.1 which results are shown in Figure 3. In Figure 3, the statistical delay bound of a session with priority level 4 (the lowest priority) when $LogPr[q_i > d] = -3$ is equal to 22 time slots.

The deterministic delay bounds of several scheduling algorithms are given in Table 1. In Table 1, N denotes the number of sessions sharing the outgoing link L_i, and ρ_i denote the maximum packet length and allocated rate for session i, L_{max} denotes the maximum packet length for all sessions except session i, r denotes the outgoing link capacity. For comparison, we assume 1 time slot = 1 second. We also assume that the bandwidth of the outgoing link $r = 1kb/s$, the packet lengths $L_i = L_{max} = 1kb$, the number of sessions $N = 41$, and the allocated rate $\rho_i = r/N = 0.024kb/s$.

Table 1 The delay bounds of scheduling algorithms

Scheduling Algorithm	Queueing Latency	Deterministic Delay Bound
WFQ [4]	$\frac{L_i}{\rho_i} - \frac{L_i}{r} + \frac{L_{max}}{r}$	41s
$SCFQ$ [5]	$\frac{L_i}{\rho_i} - \frac{L_i}{r} + (N-1)\frac{L_{max}}{r}$	80s
SFQ [7]	$(N-1)\frac{L_{max}}{r}$	40s

As shown in Table 1, the deterministic queueing delay is much larger than the statistical delay bounds. The statistical delay guarantees do not care about a small fraction of packets (e.g. one in a million packets) which experience the delay exceed the bounds. The real-time services generally can tolerate a small number of packet losses, therefore statistical delay guarantees are sufficient and suitable for these applications.

5 Conclusion

In this paper, we analyse the statistical access delay of different classes of real-time CBR services in wireless networks. Numerical results from both mathematical and simulation studies are provided. Identical results are shown from the two different methods. The results also show that the performance can be effectively differentiated by P-FCFS scheduling algorithm. The deterministic delay bounds/latencies are generally much larger than the statistical delay bounds. As real-time services generally tolerate a small number of packet losses, the statistical delay guarantees are sufficient and thus more important for real-time services. The analysis not only can provide accurate QoS performance for multiple-class real-time services but also can be used to design efficient admission control and upstream scheduling mechanisms in wireless access systems.

References

1. ITU Study Group 13, Traffic Control and Congestion Control in B-ISDN, Draft Recommendation I.371 (May 1996)
2. ATM Forum, ATM User-network Interface Specification, Version 3.1 (September 1994)
3. ATM Forum, Traffic Management Specification, Version 4.0 (February 1996)
4. Parekh, A.K., Gallerger, R.G.: A Generalized Processor Sharing Approach to Flow Control in Integrated Services Networks: The Single Node Casde. IEEE/ACM Transaction on Networking 1(3) (1993)
5. Golestani, S.J.: A Self-Clocked Fair Queueing Scheme for Broadband Applications. In: Proceedings of IEEE INFOCOM (1994)
6. Stiliadis, D., Varma, A.: Latency-Rate Servers: A General Model for Analysis of Traffic Scheduling Algorithms. IEEE/ACM Transactions on Networking 6(5) (1998)
7. Goyal, P., Vin, H.M., Cheng, H.: Start-Time Fair Queueing: A Scheduling Algorithm for Integrated Services Packet Switching Networks Technical Report TR-96-02, Department of Computer Sciences, The University of Texas at Austin (January 1996)
8. Lu, S., Raghunathan, V., Srikant, R.: Fair Scheduling in Wireless Packet Networks. IEEE/ACM Trans. Network 7(4), 473–489 (1999)
9. Raghunathan, V., Ganeriwal, S., Srivastava, M., Schurgers, C.: Energy Efficient Wireless Packet Scheduling and Fair Queueing. ACM Trans. Embedded Comput. Sys. 391, 3–23 (2004)
10. Mercankosk, G.: Mathematical Preliminaries for Rate Enforced ATM Access, Technical Memorandum of Australian Telecommunications Research Institute (ATRI), NRL-TM-071 (July 1995)
11. Roberts, J.W.: Performance evaluation and Design of Multiserverce Networks, Final Report of Management Committee of COST 224 Project (1992)
12. Humblet, P., Bhargava, A., Hluchyj, M.G.: Ballot Theorems Applied to the Transient Analysis of nD/D/1 Queues. IEEE/ACM Transactions on Networking 1(1) (1993)
13. Mercankosk, G., Budrikis, Z.L.: Establishing a Real-Time VBR Connection over an ATM Network. In: IEEE GLOBECOM 1996, pp. 1710–1714 (1996)
14. Mercankosk, G.: Access Delay Analysis for Extended Distributed Queueing Technical Memorandum of Australian Telecommunications Research Institute (ATRI), NRL-TM-022 (July 1995)

15. Zhou, H.: Real-Time Services over High Speed Networks Ph.D thesis, Curtain University of Technology (2002)
16. Iida, K., Takine, T., Sunahara, H., Oie, Y.: Delay Analysis for CBR Traffic Under Static-Priority Scheduling. IEEE/ACM Transactions on Networking 9(2) (April 2001)
17. Wongthavarawat, K., Ganz, A.: Packet Scheduling for QoS Support in IEEE 802.16 Broadband Wireless Access Systems. International Journal of Communication Systems (2003)
18. IEEE 802.16 Standard, Local and Metropolitan Area Networks, Part 16 IEEE Draft P802.16-14
19. Georgiadis, L., Guerin, R., Parekh, A.: Optimal Multiplexing on A Single Link: Delay and Buffer Requirements. In: Proceedings of IEEE INFOCOM 1994, vol. 2 (1994)

Sin and Sigmoid Higher Order Neural Networks for Data Simulations and Predictions

Ming Zhang

Abstract. New open box and nonlinear model of Sin and Sigmoid Higher Order Neural Network (SS-HONN) is presented in this paper. A new learning algorithm for SS-HONN is also developed from this study. A time series data simulation and analysis system, SS-HONN Simulator, is built based on the SS-HONN models too. Test results show that every error of SS-HONN models are from 2.1767% to 4.3114%, and the average error of Polynomial Higher Order Neural Network (PHONN), Trigonometric Higher Order Neural Network (THONN), and Sigmoid polynomial Higher Order Neural Network (SPHONN) models are from 2.8128% to 4.9076%. It means that SS-HONN models are 0.1131% to 0.6586% better than PHONN, THONN, and SPHONN models.

1 Introduction and Motivations

Many studies use traditional artificial neural network models. Blum and Li [2] studied approximation by feed-forward networks. Gorr [4] studied the forecasting behavior of multivariate time series using neural networks. Barron, Gilstrap, and Shrier [1] used polynomial neural networks for the analogies and engineering applications. However, all of the studies above use traditional artificial neural network models - black box models that did not provide users with a function that describe the relationship between the input and output. The first motivation of this paper is to develop nonlinear "open box" neural network models that will provide rationale for network's decisions, also provide better results.

Onwubolu [9] described real world problems of nonlinear and chaotic processes, which make them hard to model and predict. This chapter first compares the neural network (NN) and the artificial higher order neural network (HONN) and then presents commonly known neural network architectures and a number of HONN architectures. The time series prediction problem is formulated as a system identification problem, where the input to the system is the past values of a time series, and its desired output is the future values of a time series. The

Ming Zhang
Department of Physics, Computer Science and Engineering
Christopher Newport University, Newport News, Virginia, USA
e-mail: mzhang@cnu.edu

R. Lee (Ed.): Software Eng., Artificial Intelligence, NPD 2011, SCI 368, pp. 41–54.
springerlink.com   © Springer-Verlag Berlin Heidelberg 2011

polynomial neural network (PNN) is then chosen as the HONN for application to the time series prediction problem. This chapter presents the application of HONN model to the nonlinear time series prediction problems of three major international currency exchange rates, as well as two key U.S. interest rates—the Federal funds rate and the yield on the 5-year U.S. Treasury note. Empirical results indicate that the proposed method is competitive with other approaches for the exchange rate problem, and can be used as a feasible solution for interest rate forecasting problem. This implies that the HONN model can be used as a feasible solution for exchange rate forecasting as well as for interest rate forecasting.

Ghazali and Al-Jumeily [3] discussed the use of two artificial Higher Order Neural Networks (HONNs) models; the Pi-Sigma Neural Networks and the Ridge Polynomial Neural Networks, in financial time series forecasting. The networks were used to forecast the upcoming trends of three noisy financial signals; the exchange rate between the US Dollar and the Euro, the exchange rate between the Japanese Yen and the Euro, and the United States 10-year government bond. In particular, we systematically investigate a method of pre-processing the signals in order to reduce the trends in them. The performance of the networks is benchmarked against the performance of Multilayer Perceptrons. From the simulation results, the predictions clearly demonstrated that HONNs models, particularly Ridge Polynomial Neural Networks generate higher profit returns with fast convergence, therefore show considerable promise as a decision making tool. It is hoped that individual investor could benefit from the use of this forecasting tool.

Sanchez, Alanis, and Rico [12] proposed the use of *Higher Order Neural Networks* (HONNs) trained with an *extended Kalman filter* based algorithm to predict the electric load demand as well as the electricity prices, with beyond a horizon of 24 hours. Due to the *chaotic behavior* of the electrical markets, it is not advisable to apply the traditional forecasting techniques used for time series; the results presented here confirm that HONNs can very well capture the complexity underlying electric load demand and electricity prices. The proposed neural network model produces very accurate next day predictions and also, prognosticates with very good accuracy, a week-ahead demand and price forecasts.

Nobel Prize in Economic in 2003 rewarded two contributions: nonstationarity and time-varying volatility. These contributions had greatly deepened our understanding of properties of many economic time series (Vetenskapsakademien [13]). Granger and Bates [5] researched the combination of forecasts. Granger and Weiss [6] showed the importance of cointegration in the modeling of nonstationary economic series. Granger and Lee [17] studied multicointegration. Granger and Swanson [8] further developed multicointegration in studying of cointegrated variables. The second motivation of this paper is to develop a new nonstationary data analysis system by using new generation computer techniques that will improve the accuracy of the data simulation.

Psaltis, Park, and Hong [10] studied higher order associative memories and their optical implementations. Redding, Kowalczyk, Downs [11] developed constructive high-order network algorithm. Zhang, Murugesan, and Sadeghi [14] developed a Polynomial Higher Order Neural Network (PHONN) model for data simulation.

The contributions of this paper will be:

- Present a new model – SS-HONN (Section 2).
- Based on the SS-HONN models, build a time series simulation system – SS-HONN simulator (Section 3).
- Develop the SS-HOHH learning algorithm and weight update formulae (Section 4).
- Shows that SS-HONN can do better than PHONN, THONN, and SPHONN models. (Section 5).

2 Models of SS-HONN

SS-HONN model structure can be seen in Figure 1. Formula (1) (2) and (3) are for SS-HONN model 1b, 1 and 0 respectively. Model 1b has three layers of weights changeable, Model 1 has two layers of weights changeable, and model 0 has one layer of weights changeable. For models 1b, 1 and 0, Z is the output while x and y are the inputs of SS-HONN. $a_{kj}{}^{o}$ is the weight for the output layer, $a_{kj}{}^{hx}$ and $a_{kj}{}^{hy}$ are the weights for the second hidden layer, and $a_k{}^{x}$ and $a_j{}^{y}$ are the weights for the first hidden layer. Functions polynomial and sigmoid are the first hidden layer nodes of SS-HONN. The nodes of the second hidden layer are multiplication neurons. The output layer node of SS-HONN is a linear function of $f^{o}(net^{o}) = net^{o}$, where net^{o} equals the input of output layer node. SS-HONN is an open neural network model, each weight of HONN has its corresponding coefficient in the model formula, and each node of SS-HONN has its corresponding function in the model formula. The structure of SS-HONN is built by a nonlinear formula. It means, after training, there is rationale for each component of SS-HONC in the nonlinear formula.

For formula 1, 2, and 3, values of k and j ranges from 0 to n, where n is an integer. The SS-HONN model can simulate high frequency time series data, when n increases to a big number. This property of the model allows it to easily simulate and predicate high frequency time series data, since both k and j increase when there is an increase in n.

Figure 1 shows the SS-HONN Architecture. This model structure is used to develop the model learning algorithm, which make sure the convergence of learning. This allows the deference between desired output and real output of SS-HONN close to zero.

Formula 1:

$SS-HONN \quad Model \quad 1b:$

$$Z = \sum_{k,j=0}^{n} (c_{kj}{}^{o})\{c_{kj}{}^{hx}\sin^{k}(c_k{}^{x}x)\}\{c_{kj}{}^{hy}(1/(1+\exp(c_j{}^{y}(-y))))^{j}\}$$

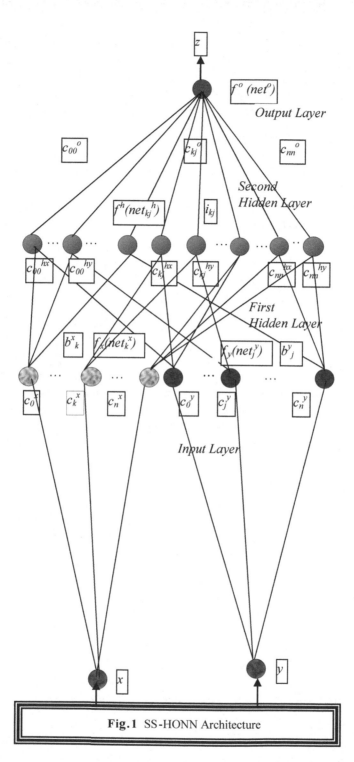

Fig.1 SS-HONN Architecture

Formula 2:

$SS-HONN \quad Model \quad 1:$

$$z = \sum_{k,j=0}^{n} c_{kj}{}^{o} \; \sin^{k}(c_{k}{}^{x}x)(1/(1+\exp(c_{j}{}^{y}(-y))))^{j}$$

$where \quad (c_{kj}{}^{hx}) = (c_{kj}{}^{hy}) = 1$

Formula 3:

$PS - HONN \qquad Model \qquad 0:$

$$z = \sum_{k,j=0}^{n} c_{kj}{}^{o} \; \sin^{k}(x)(1/(1+\exp(-y)))^{j}$$

$where : \qquad (c_{kj}{}^{hx}) = (c_{kj}{}^{hy}) = 1$

$and \qquad c_{k}{}^{x} = c_{j}{}^{y} = 1$

3 SS-HONN Time Series Analysis System

The SS-HONN simulator is written in C language, runs under X window on Sun workstation, based on previous work by Zhang, Fulcher, Scofield [11]. A user-friendly *GUI* (Graphical User Interface) system has also been incorporated. When you run the system, any step, data or calculation can be reviewed and modified

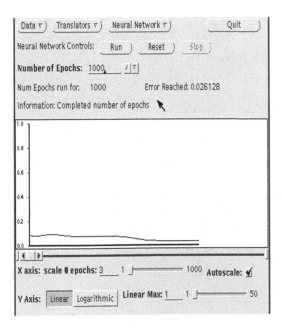

Fig. 2 SS-HONN Simulator

from different windows during processing. Hence, changing data, network models and comparing results can be done very easily and efficiently. SS-HONN simulator GUI is shown in Figure 2.

4 Learning Algorithm of SS-HONN

Learning Algorithm of SS-HONN Model can be described as followings.

The 1st hidden layer weights are updated according to:

$$c_k^x(t+1) = c_k^x(t) - \eta(\partial E_p / \partial c_k^x)$$

$$c_j^y(t+1) = c_j^y(t) - \eta(\partial E_p / \partial c_j^y)$$

Where:

$c_k^x = 1^{st}$ hidden layer weight for input x
$k = k$th neuron of first hidden layer
$c_j^y = 1^{st}$ hidden layer weight for input y
$j = j$th neuron of first hidden layer
η = learning rate (positive & usually < 1)
E_p = error
t = training time

The equations for the kth or jth node in the first hidden layer are:

$$net_k^x = c_k^x * x$$

$$b_k^x = f_x(net_k^x) = Sin^k(c_k^x * x)$$

or

$$net_j^y = c_j^y * (-y)$$

$$b_j^y = f_y(net_j^y) = (1/(1+\exp(c_j^y *(-y))))^j$$

Where:

b_k^x and b_j^y = output from the 1st hidden layer neuron (= input to 2nd hidden layer neuron)
f_x and $f_y = 1^{st}$ hidden layer neuron activation function
x and y = input to 1st hidden layer

$$net_{kj}^h = c_{kj}^{hx} b_k^x * c_{kj}^{hy} b_j^y$$

$$i_{kj} = f^h(net_{kj}^h) = c_{kj}^{hx} b_k^x * c_{kj}^{hy} b_j^y$$

Where:

i_{kj} = output from 2nd hidden layer (= input to the output neuron)

$$net^o = \sum_{k,j=0}^{n} c_{kj}{}^o i_{kj}$$

$$z = f^o(net^o) = \sum_{k,j=0}^{n} c_{kj}{}^o i_{kj}$$

Where:

Z is output of HONN.

The total error is the sum of the squared errors across all hidden units, namely:

$$E_p = 0.5 * \delta^2 = 0.5 * (d-z)^2$$
$$= 0.5 * (d - f^o(net^o))^2$$
$$= 0.5 * (d - f^o(\sum_j c_{kj}{}^o i_{kj}))^2$$

Where:

d is actual output of HONN

For polynomial function and sigmoid function as in the first layer of HONN:

$$b^x{}_k = f_x(net_k{}^x) = \sin^k(k * net_k{}^x)$$
$$f_x{}'(net_k{}^x) = \partial b^x{}_k / \partial(net_k{}^x)$$
$$= \partial(\sin^k(k * net_k{}^x)) / \partial(net_k{}^x)$$
$$= k\sin^{k-1}(k * net_k{}^x) * (\cos(k * net_k{}^x)) * k$$
$$= k^2 \sin^{k-1}(k * net_k{}^x)\cos(k * net_k{}^x)$$

$$b^y{}_j = f_y(net_j{}^y) = (1/(1+\exp(net_j{}^y)))^j$$
$$f_y{}'(net_j{}^y) = \partial b^y{}_j / \partial(net_j{}^y)$$
$$= \partial(1/(1+\exp(net_j{}^y)))^j / \partial(net_j{}^y)$$
$$= -j * \exp(net_j{}^y) * (1/(1+\exp(net_j{}^y)))^{j+1}$$

The gradient ($\partial E_p / \partial c_k^{\,x}$) is given by:

$$\partial E_p / \partial c_k^{\,x} = \partial(0.5*(d-z)^2)/\partial c_k^{\,x}$$

$$= (\partial(0.5*(d-z)^2)/\partial z)(\partial z/\partial(net^o))$$

$$(\partial(net^o)/\partial i_{kj})(\partial i_{kj}/\partial(net_{kj}^{\,h}))(\partial(net_{kj}^{\,h})/\partial b^x{}_k)$$

$$(\partial b^x{}_k / \partial(net_k^{\,x}))(\partial(net_k^{\,x})/\partial c_k^{\,x})$$

$$\partial(0.5*(d-z)^2/\partial z = -(d-z)$$

$$\partial z/\partial(net^o) = \partial f^o/\partial(net^o) = f^{o\,\prime}(net^o)$$

$$\partial(net^o)/\partial i_{kj} = \partial(\sum_{k,j=1}^{L}(c_{kj}^{\,o}i_{kj})/\partial i_{kj} = c_{kj}^{\,o}$$

$$\partial i_{kj}/\partial(net_{kj}^{\,h})$$
$$= \partial(f^{\,h}(net_{kj}^{\,h}))/\partial(net_{kj}^{\,h}) = f^{\,h\,\prime}(net_{kj}^{\,h})$$

$$\partial net_{kj}^{\,h}/\partial b^x{}_k$$
$$= \partial((c_{kj}^{\,hx}*b^x{}_k)*(c_{kj}^{\,hy}*b^y{}_j))/\partial b^x{}_k = c_{kj}^{\,hx}*c_{kj}^{\,hy}*b^y{}_j$$
$$= \delta_{kj}^{\,hx}c_{kj}^{\,hx}$$
$$where \, \delta_{kj}^{\,hx} = c_{kj}^{\,hy}*b^y{}_j$$

$$\partial b^x{}_k / \partial(net_k^{\,x}) = f_x{}'(net_k^{\,x})$$

$$\partial(net_k^{\,x})/\partial c_k^{\,x} = \partial(c_k^{\,x}*x)/\partial c_k^{\,x} = x$$

Combining Formulae the negative gradient is:

$$-\partial E_p / \partial c_k^{\,x}$$
$$= (d-z)f^{o\,\prime}(net^o)c_{kj}^{\,o}*f^{\,h\,\prime}(net_{kj}^{\,h})\delta_{kj}^{\,hx}c_{kj}^{\,hx}f_x{}'(net_k^{\,x})x$$

The weight update equations are calculated as follows.
For linear output neurons:

$$f^{o\prime}(net^o) = 1$$

$$\delta^{ol} = (d - z)f^{o\prime}(net^o) = (d - z)$$

For linear neurons of second hidden layer:

$$f^{h\prime}(net_{kj}^{h}) = 1$$

The negative gradient is:

$$-\partial E_p / \partial c_k^{x}$$

$$= (d-z)f^{o\prime}(net^o)c_{kj}^{o} * f^{h\prime}(net_{kj}^{h})\delta_{kj}^{hx}c_{kj}^{hx}f_x^{\prime}(net_k^{x})x$$

$$= \delta^{ol} * c_{kj}^{o} * \delta_{kj}^{hx} * c_{kj}^{hx} * k * (net_k^{x})^{k-1} * x$$

By combining above formulae, the learning algorithm for the 1st hidden layer weights of x input neuron:

$$c_k^{x}(t+1) = c_k^{x}(t) - \eta(\partial E_p / \partial c_k^{x})$$

$$= c_k^{x}(t) + \eta(d-z)f^{o\prime}(net^o)c_{kj}^{o} * f^{h\prime}(net_{kj}^{h})\delta_{kj}^{hx}c_{kj}^{hx}f_x^{\prime}(net_k^{x})x$$

$$= c_k^{x}(t) +$$

$$\eta * \delta^{ol} * c_{kj}^{o} * \delta^{hx} * c_{kj}^{hx} * (k^2)\sin^{k-1}(k*net_k^{x})\cos(k*net_k^{x}) * x$$

$$= c_k^{x}(t) + \eta * \delta^{ol} * c_{kj}^{o} * \delta^{hx} * c_{kj}^{hx} * \delta^{x} * x$$

where:

$$\delta^{ol} = (d-z)f^{o\prime}(net^o) = d - z \qquad (linear)$$

$$\delta^{hx} = f^{h\prime}(net_{kj}^{h})c_{kj}^{hy}b^y_j \quad = c_{kj}^{hy}b_j \qquad (linear)$$

$$\delta^{x} = f_x^{\prime}(net_k^{x}) = (k^2)\sin^{k-1}(k*net_k^{x})\cos(k*net_k^{x})$$

Using the above procedure, the learning algorithm for the 1st hidden layer weights of y input neuron:

$$c_j^{y}(t+1) = c_j^{y}(t) - \eta(\partial E_p / \partial c_j^{y})$$

$$= c_j^{y}(t) + \eta(d - z)f^{o\prime}(net^o)c_{kj}^{o} * f^{h\prime}(net_{kj}^{h})\delta_{kj}^{hy}c_{kj}^{hy}f_y^{\prime}(net_j^{y})y$$

$$= c_j^{y}(t) + \eta * \delta^{ol} * c_{kj}^{o} * \delta^{hy} * c_{kj}^{hy}$$

$$* (-j) * \exp(net_j^{y}) * (1/(1 + \exp(net_j^{y})))^{j+1} * y$$

$$= c_j^{y}(t) + \eta * \delta^{ol} * c_{kj}^{o} * \delta^{hy} * c_{kj}^{hy} * \delta^{y} * y$$

where :

$$\delta^{ol} = (d-z)f^{o}{}'(net^{o}) = d-z \qquad (linear)$$

$$\delta^{hy} = f^{h}{}'(net_{kj}^{hy})c_{kj}^{hx}b^{x}{}_{k} \quad = c_{kj}^{hx}b^{x}{}_{k} \qquad (linear)$$

$$\delta^{y} = f_{y}{}'(net_{j}^{y}) = (-j)*\exp(net_{j}^{y})*(1/(1+\exp(net_{j}^{y})))^{j+1}$$

5 Time Series Data Test Using SS-HONN

This paper uses the monthly Switzerland Franc and USA dollar exchange rate from November 2008 to December 2009 (See Table 1 and 2) as the test data for SS-HONN models. This paper also uses the monthly New Zealand dollar and USA dollar exchange rate from November 2008 to December 2009 (See Table 3 and 4) as the test data for SS-HONN models. Rate and desired output data, R_t, are from USA Federal Reserve Bank Data bank. Input1, R_{t-2}, are the data at time t-2. Input 2, R_{t-1} are the data at time t-1.The values of R_{t-2}, R_{t-1}, and R_t are converted to a range from 0 to 1 and then used as inputs and output in the SS-HONN model. SS-HONN model 1b is used for Table 1, 2, 3 and 4. The test data of SS-HONN orders 6 for using 10,000 epochs are shown on the tables.

In Table 1, "SwitzelandFranc/USDollar Exchange Rate CHF\$1.00 = USD\$0.8561 on 3-Nov-08", the average errors of PHONN, THONN, SPHONN, and SS-HONN are 2.7552%, 2.8547%, 2.8286%, and 2.6997% respectively. The average error of PHONN, THONN, and SPHONN is 2.8128%. So SS-HONN error is 0.1131% better than the average error of PHONN, THONN, and SPHONN models.

In Table 2, "SwitzelandFranc/USDollar Exchange Rate CHF\$1.00 = USD\$0.8220 on 28-Nov-08", the average errors of PHONN, THONN, SPHONN, and SS-HONN are 2.1962%, 3.5549%, 2.7549%, and 2.1767% respectively. The average error of PHONN, THONN, and SPHONN is 2.8353%. So SS-HONN error is 0.6586% better than the average error of PHONN, THONN, and SPHONN models.

In Table 3, "NewZealandDollar/USDollar Exchange RateNZD\$1.00 = USD\$0.5975 on 3-Nov-09", the average errors of PHONN, THONN, SPHONN, and SS-HONN are 4.3653%, 5.6771%, 4.6806%, and 4.3113% respectively. The average error of PHONN, THONN, and SPHONN is 4.9076%. So SS-HONN error is 0.5962% better than the average error of PHONN, THONN, and SPHONN models.

In Table 4, "NewZealandDollar/USDollar Exchange RateNZD\$1.00 = USD\$0.5500 on 28-Nov-08", the average errors of PHONN, THONN, SPHONN, and SS-HONN are 4.2512%, 4.6730%, 5.19452%, and 4.2354% respectively. The average error of PHONN, THONN, and SPHONN is 4.7062%. So SS-HONN error is 0.4708% better than the average error of PHONN, THONN, and SPHONN models.

Table 1 SwitzelandFranc/USDollar Exchange Rate

CHF$1.00 = USD$0.8561 on 3-Nov-08 , USA Federal Reserve Bank Data (CHF-USD2009-1.dat)

Original Data				HONN Output				HONN Error (% Percentage)			
Date	Rate Desired Output	Input1 2 month ago data	Input2 1 month ago data	PHONN	THONN	SPHONN	SS-HONN	PHONN	THONN	SPHONN	SS-HONN
11/3/08	0.8561										
12/1/08	0.8281										
1/2/09	0.9359	0.8561	0.8281	0.8812	0.8769	0.8755	0.8730	5.8430	6.3031	6.4530	6.7205
2/2/09	0.8591	0.8281	0.9359	0.8946	0.9074	0.8690	0.9076	4.1348	5.6271	1.1502	5.6425
3/2/09	0.8530	0.9359	0.8591	0.9317	0.9262	0.9475	0.8997	9.2218	8.5760	11.0770	5.4762
4/1/09	0.8722	0.8591	0.8530	0.8904	0.8874	0.8818	0.8890	2.0869	1.7424	1.0994	1.9262
5/1/09	0.8800	0.8530	0.8722	0.8928	0.8921	0.8797	0.8999	1.4608	1.3811	0.0303	2.2689
6/1/09	0.9356	0.8722	0.8800	0.9062	0.9030	0.8967	0.9062	3.1402	3.4828	4.1572	3.1402
7/1/09	0.9328	0.8800	0.9356	0.9292	0.9269	0.9115	0.9371	0.3914	0.6384	2.2919	0.4568
8/3/09	0.9454	0.9356	0.9328	0.9619	0.9518	0.9581	0.9466	1.7537	0.6836	1.3511	0.1326
9/1/09	0.9387	0.9328	0.9454	0.9659	0.9548	0.9576	0.9533	2.9019	1.7175	2.0163	1.5575
10/1/09	0.9593	0.9454	0.9387	0.9704	0.9593	0.9678	0.9522	1.1493	0.0008	0.8883	0.7405
11/2/09	0.9826	0.9387	0.9593	0.9754	0.9624	0.9645	0.9618	0.7378	2.0527	1.8387	2.1139
12/1/09	1.0016	0.9593	0.9826	0.9992	0.9811	0.9857	0.9794	0.2403	2.0503	1.5903	2.2203
Average Error (% Percentage)								2.7552	2.8547	2.8286	2.6997
Average Error pf PHONN, THONN, and SPHONN(% Percentage)								2.8128	SS-HONN Better		0.1131

Table 2 SwitzelandFranc/USDollar Exchange Rate

CHF$1.00 = USD$0.8220 on 28-Nov-08, USA Federal Reserve Bank Data (CHF-USD2009-2.dat)

Original Data				HONN Output				HONN Error (% Percentage)			
Date	Rate Desired Output	Input1 2 month ago data	Input2 1 month ago data	PHONN	THONN	SPHONN	SS-HONN	PHONN	THONN	SPHONN	SS-HONN
11/28/08	0.8220										
12/31/08	0.9369										
1/30/09	0.8612	0.8220	0.9369	0.8811	0.9418	0.9446	0.9066	2.3101	9.3580	9.6816	5.2794
2/27/09	0.8568	0.9369	0.8612	0.9445	0.9396	0.8796	0.8868	10.237	9.6676	2.6646	3.5124
3/31/09	0.8776	0.8612	0.8568	0.8629	0.8718	0.8822	0.8744	1.6766	0.6562	0.5229	0.3614
4/30/09	0.8770	0.8568	0.8776	0.8701	0.8805	0.8978	0.8890	0.7881	0.3918	2.3658	1.3675
5/29/09	0.9353	0.8776	0.8770	0.8913	0.8790	0.8958	0.8912	4.7000	6.0191	4.2212	4.7106
6/30/09	0.9202	0.8770	0.9353	0.9249	0.9035	0.9390	0.9322	0.5037	1.8210	2.0391	1.3039
7/31/09	0.9374	0.9353	0.9202	0.9553	0.9314	0.9224	0.9287	1.9114	0.6362	1.6021	0.9228
8/31/09	0.9462	0.9202	0.9374	0.9531	0.9173	0.9365	0.9388	0.7342	3.0516	1.0219	0.7801
9/30/09	0.9639	0.9374	0.9462	0.9636	0.9369	0.9412	0.9467	0.0309	2.7975	2.3536	1.7755
10/30/09	0.9768	0.9462	0.9639	0.9708	0.9511	0.9532	0.9596	0.6073	2.6243	2.4104	1.7584
11/30/09	0.9950	0.9639	0.9768	0.9739	0.9799	0.9607	0.9700	2.1224	1.5225	3.4524	2.5124
12/31/09	0.9654	0.9768	0.9950	0.9725	1.0051	0.9724	0.9832	0.7329	4.1133	0.7226	1.8357
Average Error (% Percentage)								2.1962	3.5549	2.7549	2.1767
Average Error pf PHONN, THONN, and SPHONN(% Percentage)								2.8353	SS-HONN Better		0.6586

Table 3 NewZealandDollar/USDollar Exchange Rate

NZD$1.00 = USD$0.5975 on 3-Nov-09, USA Federal Reserve Bank Data (NZD-USD2009-1.dat)

Original Data				HONN Output				HONN Error (% Percentage)			
Date	Rate Desired Output	Input1 2 month ago data	Input2 1 month ago data	PHONN	THONN	SPHONN	SS-HONN	PHONN	THONN	SPHONN	SS-HONN
11/3/08	0.5975										
12/1/08	0.5355										
1/2/09	0.5850	0.5975	0.5355	0.5706	0.5599	0.5914	0.5746	2.4682	4.2988	1.0934	1.7708
2/2/09	0.5026	0.5355	0.5850	0.5883	0.5867	0.5934	0.5829	17.0441	16.7252	18.0587	15.9821
3/2/09	0.4926	0.5850	0.5026	0.5486	0.5413	0.5668	0.5503	11.3602	9.8961	15.0575	11.7152
4/1/09	0.5635	0.5026	0.4926	0.5310	0.5474	0.5290	0.5245	5.7669	2.8580	6.1159	6.9175
5/1/09	0.5687	0.4926	0.5635	0.5629	0.5906	0.5649	0.5670	1.0178	3.8500	0.6719	0.3004
6/1/09	0.6509	0.5635	0.5687	0.5856	0.5745	0.5964	0.5889	10.0259	11.7383	8.3807	9.5223
7/1/09	0.6452	0.5687	0.6509	0.6449	0.6281	0.6377	0.6484	0.0401	2.6484	1.1580	0.4905
8/3/09	0.6683	0.6509	0.6452	0.6648	0.6484	0.6703	0.6649	0.5194	2.9829	0.2982	0.5085
9/1/09	0.6794	0.6452	0.6683	0.6825	0.6652	0.6780	0.6801	0.4502	2.0911	0.2039	0.1071
10/1/09	0.7168	0.6683	0.6794	0.6954	0.6877	0.6917	0.6932	2.9920	4.0592	3.5002	3.2969
11/2/09	0.7225	0.6794	0.7168	0.7271	0.7343	0.7110	0.7239	0.6389	1.6371	1.5895	0.1952
12/1/09	0.7285	0.7168	0.7225	0.7289	0.7674	0.7288	0.7353	0.0600	5.3400	0.0400	0.9300
Average Error (% Percentage)								4.3653	5.6771	4.6806	4.3114
Average Error pf PHONN, THONN, and SPHONN(% Percentage)								4.9076	SS-HONN Better		0.5962

Table 4 NewZealandDollar/USDollar Exchange Rate

NZD$1.00 = USD$0.5500 on 28-Nov-08, USA Federal Reserve Bank Data (NZD-USD2009-2.dat)

Original Data				HONN Output				HONN Error (% Percentage)			
Date	Rate Desired Output	Input1 2 month ago data	Input2 1 month ago	PHONN	THONN	SPHONN	SS-HONN	PHONN	THONN	SPHONN	SS-HONN
11/28/08	0.5500										
12/31/08	0.5815										
1/30/09	0.5084	0.5500	0.5815	0.5911	0.5974	0.5882	0.5582	16.2598	17.5013	15.6890	9.8024
2/27/09	0.5030	0.5815	0.5084	0.5390	0.5318	0.5655	0.5196	7.1663	5.7240	12.4309	3.2911
3/31/09	0.5692	0.5084	0.5030	0.5293	0.5269	0.5376	0.5446	7.0055	7.4388	5.5524	4.3288
4/30/09	0.5695	0.5030	0.5692	0.5758	0.5858	0.5638	0.5922	1.1114	2.8694	1.0033	3.9905
5/29/09	0.6370	0.5692	0.5695	0.5851	0.5860	0.5903	0.5693	8.1451	8.0084	7.3365	10.6280
6/30/09	0.6447	0.5695	0.6370	0.6400	0.6509	0.6221	0.6536	0.7347	0.9646	3.5030	1.3809
7/31/09	0.6605	0.6370	0.6447	0.6646	0.6572	0.6565	0.6571	0.6254	0.4950	0.6049	0.5170
8/31/09	0.6856	0.6447	0.6605	0.6792	0.6722	0.6681	0.6700	0.9316	1.9474	2.5506	2.2755
9/30/09	0.7233	0.6605	0.6856	0.7016	0.6958	0.6886	0.6860	3.0059	3.7983	4.8013	5.1524
10/30/09	0.7230	0.6856	0.7233	0.7265	0.7297	0.7206	0.7092	0.4863	0.9278	0.3366	1.9020
11/30/09	0.7151	0.7233	0.7230	0.7243	0.7275	0.7389	0.7388	1.2920	1.7283	3.3312	3.3211
12/31/09	0.7255	0.7230	0.7151	0.7227	0.7203	0.7349	0.7352	0.3900	0.7100	1.2900	1.3400
Average Error (% Percentage)								4.2513	4.6730	5.1945	4.2354
Average Error pf PHONN, THONN, and SPHONN(% Percentage)								4.7062	SS-HONN Better		0.4708

6 Conclusion

This paper develops the details of a open box and nonlinear higher order neural network models of SS-HONN. This paper also provides the learning algorithm formulae for SS-HONN, based on the structures of SS-HONN. This paper uses SS-HONN simulator and tests the SS-HONN models using high frequency data and the running results are compared with Polynomial Higher Order Neural Network (PHONN), Trigonometric Higher Order Neural Network (THONN), and Sigmoid polynomial Higher Order Neural Network (SPHONN) models. Test results show that every error of SS-HONN models are from 2.1767% to 4.3114%, and the average error of Polynomial Higher Order Neural Network (PHONN), Trigonometric Higher Order Neural Network (THONN), and Sigmoid polynomial Higher Order Neural Network (SPHONN) models are from 2.8128% to 4.9076%. It means that SS-HONN models are 0.1131% to 0.6586% better than PHONN, THONN, and SPHONN models. One of the topics for future research is to continuebuilding models of higher order neural networks for different data series. The coefficients of the higher order models will be studied not only using artificial neural network techniques, but also statistical methods.

Acknowledgments

The author would like to acknowledge the financial assistance of the following organizations in the development of Higher Order Neural Networks: Fujitsu Research Laboratories, Japan (1995-1996), Australian Research Council (1997-1998), the US National Research Council (1999-2000), and the Applied Research Centers and Dean's Office Grants of our University, USA (2001-2010).

References

1. Barron, R., Gilstrap, L., Shrier, S.: Polynomial and Neural Networks: Analogies and Engineering Applications. In: Proceedings of International Conference of Neural Networks, New York, vol. 2, pp. 431–439 (1987)
2. Blum, E., Li, K.: Approximation theory and feed-forward networks. Neural Networks 4, 511–551 (1991)
3. Ghazali, R., Al-Jumeily, D.: Application of pi-sigma neural networks and ridge polynomial neural networks to financial time series prediction. In: Zhang, M. (ed.) Artificial Higher Order Neural Networks for Economics and Business, pp. 190–211. Information Science Reference (an imprint of IGI Global), Hershey (2009)
4. Gorr, W.L.: Research prospective on neural network forecasting. International Journal of Forecasting 10(1), 1–4 (1994)
5. Granger, C.W.J.: Some properties of time series data and their use in econometric model specification. Journal of Econometrics 16, 121–130 (1981)
6. Granger, C.W.J., Weiss, A.A.: Time series analysis of error-correction models. In: Karlin, S., Amemiya, T., Goodman, L.A. (eds.) Studies in Econometrics. Time Series and Multivariate Statistics, pp. 255–278. Academic Press, San Diego (1983), In Honor of T. W. Anderson

7. Granger, C.W.J., Lee, T.H.: Multicointegration. In: Rhodes Jr., G.F., Fomby, T.B. (eds.) Advances in Econometrics: Cointegration, Spurious Regressions and Unit Roots, pp. 17–84. JAI Press, New York (1990)
8. Granger, C.W.J., Swanson, N.R.: Further developments in study of cointegrated variables. Oxford Bulletin of Economics and Statistics 58, 374–386 (1996)
9. Onwubolu, G.C.: Artificial higher order neural networks in time series prediction. In: Zhang, M. (ed.) Artificial Higher Order Neural Networks for Economics and Business, pp. 250–270. Information Science Reference (an imprint of IGI Global), Hershey (2009)
10. Psaltis, D., Park, C., Hong, J.: Higher Order Associative Memories and their Optical Implementations. Neural Networks 1, 149–163 (1988)
11. Redding, N., Kowalczyk, A., Downs, T.: Constructive high-order network algorithm that is polynomial time. Neural Networks 6, 997–1010 (1993)
12. Selviah, D.R., Shawash, J.: Generalized correlation higher order neural networks for financial time series prediction. In: Zhang, M. (ed.) Artificial Higher Order Neural Networks for Economics and Business, pp. 212–249. Information Science Reference (an imprint of IGI Global), Hershey (2009)
13. Vetenskapsakademien, K.: Time-series econometrics: Co-integration and Autoregressive Conditional Heteroskedasticity. In: Advanced Information on the Bank of Sweden Prize in Economic Sciences in Memory of Alfred Nobel, pp. 1–20 (2003)
14. Zhang, M., Murugesan, S., Sadeghi, M.: Polynomial higher order neural network for economic data simulation. In: Proceedings of International Conference on Neural Information Processing, Beijing, China, pp. 493–496 (1995)

A Proposal of Mouth Shapes Sequence Code for Japanese Pronunciation

Tsuyoshi Miyazaki, Toyoshiro Nakashima, and Naohiro Ishii

Abstract. In this paper, we examine a method in which distinctive mouth shapes are processed using a computer. When lip-reading skill holders do lip-reading, they stare at the changes in mouth shape of a speaker. In recent years, some researches into lip-reading using information technology has been pursued. There are some researches based on the changes in mouth shape. The researchers analyze all data of the mouth shapes during an utterance, whereas lip-reading skill holders look at distinctive mouth shapes. We found that there was a high possibility for lip-reading by using the distinctive mouth shapes. To build the technique into a lip-reading system, we propose an expression method of the distinctive mouth shapes which can be processed using a computer. In this way, we acquire knowledge about the relation between Japanese phones and mouth shapes. We also propose a method to express order of the distinctive mouth shapes which are formed by a speaker.

1 Introduction

In recent years, some researches for realizing lip-reading with information process-ing technologies have been pursued. This is called "machine lip-reading". It is used as a complimentary technology to improve speech recognition, and the research of lip-reading for supporting communication with hearing-impaired people is being pursued.

Tsuyoshi Miyazaki
Kanagawa Institute of Technology, 1030 Shimo-ogino, Atsugi, Kanagawa, Japan
e-mail: miyazaki@ic.kanagawa-it.ac.jp

Toyoshiro Nakashima
Sugiyama Jogakuen University, 17-3 Hoshigaoka-motomachi, Chikusa, Nagoya, Aichi, Japan
e-mail: nakasima@sugiyama-u.ac.jp

Naohiro Ishii
Aichi Institute of Technology, 1247 Yachigusa, Yakusa, Toyota, Aichi, Japan
e-mail: ishii@aitech.ac.jp

R. Lee (Ed.): Software Eng., Artificial Intelligence, NPD 2011, SCI 368, pp. 55–65.
springerlink.com   © Springer-Verlag Berlin Heidelberg 2011

In the machine lip-reading, using a camera, take several images (frame images) of the lips region when a speaker is uttering. Then, digital image processing for the frame images is computed to extract the changes in mouth shape and the sequential movements of lips during the utterance. Next, features are computed from the results and an utterance word or phrase is inferred by the features[1, 2, 3, 4, 5, 6, 7].

For example, a method using optical flow generates the features from direction and distance of the movement between the images of the feature points around the lips region[1, 2]. In addition, a method using mouth shape changes generates the features from aspect ratio of the lips region[3, 4]. A method using images of the lips region generates the features from sum of the difference of pixel value between the frame images by template matching[5].

On the other hand, we found that people who have acquired a skill of lip-reading ("lip-reading skill holder" is used hereafter) stare at the mouth shape of a speaker when reading lips. In addition, some distinctive mouth shapes are formed intermittently when uttering Japanese.

The methods proposed by Li[3] and Saito[4] are similar to the method of lip-reading skill holders in which it stares at mouth shapes. However, their methods focus on a sequence of mouth shape changes from the start to the end of utterance, whereas the lip-reading skill holders stare at distinctive mouth shapes appearing intermittently.

In this study, as a first step for realizing the machine lip-reading by modeling the lip-reading skill holder, we examine a method in which knowledge of the lip-reading skill holders is logically materialized and the distinctive mouth shapes are processed using a computer. Based on this principle, we propose a method to express the distinctive mouth shapes which are formed sequentially when people produce utterance of an arbitrary word or phrase.

2 Relation between Japanese Phones and Mouth Shapes

We explain the relation between Japanese phones and mouth shapes. In addition, the length of voice equivalent of one short syllable is called "mora", and the voice is called "phone".

The mouth shapes become to that of the vowels when uttering the Japanese phones. There is a Japanese syllables are usually represented as a table that is called "Japanese syllabary table" in Japanese(Table 1). In this table, the vowels assume the row and the consonants assume the column. Each phone in Japanese is arranged in this table. Therefore, it is easy to recognize the phones that are formed the same mouth shape because the phone of each row is the same vowel. However, there are some specific phones with which mouth shape forms are different at the beginning of utterance. For instance, a Japanese phone, "sa" forms the mouth shape of /i/ at the beginning of the utterance, and it forms the mouth shape of /a/ afterwards. In the same manner, a Japanese phone, "wa" forms mouth shape of /u/ at the beginning, and it forms mouth shape of /a/ afterward. Such mouth shapes in Japanese

Table 1 The Japanese syllabary table

	/k/ /s/ /t/ /n/ /h/ /m/ /y/ /r/ /w/
/a/	a ka sa ta na ha ma ya ra wa
/i/	i ki si ti ni hi mi — ri —
/u/	u ku su tu nu hu mu yu ru —
/e/	e ke se te ne he me — re —
/o/	o ko so to no ho mo yo ro wo

phone consist of three kinds of types including the closed mouth shape. These mouth shapes are necessary to utter given Japanese phone correctly.

As just described, although the mouth shapes formed to utter all the Japanese phones are not only those of vowel, all the Japanese phones can be uttered by combination of the mouth shapes /a/, /i/, /u/, /e/, /o/ and the closed mouth shape. And these six mouth shapes shall be defined as "Basic Mouth Shape" when uttering Japanese phones, and the combinations of these six shall be able to be classified 16 patterns. The 16 mouth shape patterns are shown in Table 2.

Table 2 The relation of Japanese phones and mouth shape patterns

#	vowel	BeMS	EMS	Japanese phones
1	/a/	–	/a/	a, ka, ha, ga
2		/i/	/a/	sa, ta, na, ya, ra, za, da, kya, sya, tya, nya, hya, rya, gya, zya
3		/u/	/a/	wa, fa
4		closed	/a/	ma, ba, pa, mya, bya, pya
5	/i/	–	/i/	i, ki, si, ti, ni, hi, ri, gi, zi, di
6		/u/	/i/	wi, fi
7		closed	/i/	mi, bi, pi
8	/u/	–	/u/	u, ku, su, tu, nu, hu, yu, ru, gu, zu, du, kyu, syu, tyu, nyu, hyu, ryu, gyu, zyu
9		closed	/u/	mu, bu, pu, myu, byu, pyu
10	/e/	–	/e/	e, ke, he, ge
11		/i/	/e/	se, te, ne, re, ze, de, kye, sye, tye, nye, hye, rye, gye, zye
12		/u/	/e/	we, fe
13		closed	/e/	me, be, pe, mye, bye, pye
14	/o/	–	/o/	o, ko, ho, go
15		/u/	/o/	so, to, no, yo, ro, wo, zo, do, kyo, syo, tyo, nyo, hyo, ryo, gyo, zyo, fo
16		closed	/o/	mo, bo, po, myo, byo, pyo

BeMS: Beginning Mouth Shape, EMS: End Mouth Shape.

From Table 2, the relation between Japanese phones and the mouth shape patterns turns out to be clear. Thus the mouth shape sequence is figured out without utterance. When two or more phones are uttered, like a word, the mouth shape patterns corresponding to the utterance phone are basically formed sequentially. But, depending on the combination of the phone, a part of the mouth shape patterns are affected from the mouth shape of other phone, and it is observed that phones may not correspond to their mouth shape patterns in some cases. For example, when uttering a Japanese word "kita", the mouth shape of /i/ which is formed at the beginning of "ta" has already been formed with the phone of "ki". Therefore, utterance of "ta" can be done without the mouth shape of /i/. Then combinations of the mouth shape patterns in case of each phone utterance are made from the pattern 5 first and then pattern 2 in Table 2, but the mouth shapes formed at actual utterance are made from combinations of the pattern 5 first and then pattern 1. In this way, when the mouth shape of the end of a phone becomes equal to the mouth shape of the beginning of the following phone, mouth shapes are not simply of combinations of the mouth shape patterns. And some other combinations of such a pattern exist there.

Furthermore, the mouth shapes that can not be included in any columns of the Japanese syllabary of geminate consonant and syllabic nasal get complicated. This is because specific mouth shapes for those phones are not definitive due to the fact that those mouth shapes are dictated by the phones uttered before and after. For example, the mouth shape for a geminate consonant and a syllabic nasal after the phone uttered with the mouth shape of /a/ will be the mouth shape of /i/.

3   Definition of Mouth Shpaes

First, mouth shapes are encoded in order to make easy handling on a computer. We call the six mouth shapes used when uttering Japanese phone as "Basic Mouth Shape" (BaMS), and define the BaMS in (1) below. Here, each symbol expresses the mouth shape /a/, /i/, /u/, /e/, /o/ and closed mouth, respectively.

$$BaMS = \{A,I,U,E,O,X\} \tag{1}$$

Next, we call the mouth shapes at the beginning of utterance of Japanese phones as "Beginning Mouth Shape" (BeMS) and the mouth shapes for vowel phones as "End Mouth Shape" (EMS). We define the BeMS and the EMS in (2) and (3) respectively.

$$BeMS = \{I,U,X\} \tag{2}$$

$$EMS = \{A,I,U,E,O,X\} \tag{3}$$

Here, we decide to call the phones uttered without forming BeMS as seen in the mouth shape patterns 1, 5, 8, 10 and 14 in Table 2 as "Simple Mouth Shape Phone", and the phones uttered with use of BeMS and EMS as "Couple Mouth Shapes Phone". In addition, we decide to call these mouth shape patterns as "Mouth Shape Segment" because each mouth shape pattern is a unit when uttering one phone.

3.1   Mouth Shapes Sequence Code

We propose an expression method of BaMS formed sequentially when uttering Japanese words or phrases. At first, we define the code for a BaMS formed at the utterance as "Mouth Shape Code" (hereinafter called MS Code) in (4) below. Here, each MS Code corresponds to each mouth shape in (1) above. But the symbol "–" is for a code corresponding to BeMS which is not formed with Simple Mouth Shape Phone.

$$C = \{A, I, U, E, O, X, -\} \tag{4}$$

Using these MS Codes, BeMS and EMS which are formed sequentially when uttering a word or phrase is expressed as follows; Citing the case when uttering Japanese "ame" (rain in English) as an example, an MS Code is defined as "–" because the BeMS is not formed when uttering the first "a" phone, and the codes "–A" is determined by connecting an MS Code for the mouth shape of /a/ formed as the EMS. In the same way as described above, a closed mouth shape is formed as the BeMS first when uttering the next phone "me". We express the codes for this Japanese as "–AXE" by adding an MS Code "X" for a closed mouth shape and "E" for a continuing mouth shape of /e/ formed as the EMS. Because this sequential codes show order of the mouth shapes formed at the time of a word or phrase utterance, we call them "Mouth Shapes Sequence Code" (MSSC). In addition, the odd numbered codes and the even numbered codes correspond to MS Codes of BeMS and EMS respectively because one Japanese phone is expressed with two MS Codes, BeMS and EMS.

Furthermore, by defining the MS Codes, we can express 16 mouth shape patterns shown in Table 2 as a pattern of MS Codes(Table 3), and the MS Codes can be assigned for all the Japanese phones. We call them "Phone Code". As the result, as shown in Section 2, it can be possible to generate an MSSC even from a word or phrase, not from utterance images by using the Phone Codes.

Table 3 The patterns of Phone Code

#	Phone Code	#	Phone Code
1	–A	9	XU
2	IA	10	–E
3	UA	11	IE
4	XA	12	UE
5	–I	13	XE
6	UI	14	–O
7	XI	15	UO
8	–U	16	XO

For example, as for the Japanese word "asahi" (morning sun in English), the Phone Code for the phone "a", "sa" and "hi" correspond to "–A", "IA" and "–I" respectively from Table 3, and by connecting them in order of utterance, the MSSC "–AIA–I" is obtained for the word "asahi". But it is necessary to pay attention when generating an MSSC from a word or phrase without actual utterance. As described in Section 2, it is because that there is a case that in actual utterance, the mouth shape order may differ from the MSSC obtained from simply connected

$$c_B(1)c_E(1)\cdots c_B(s-1)c_E(s-1)c_B(s)c_E(s)c_B(s+1)c_E(s+1)\cdots$$
$$\cdots c_B(\text{sMAX})c_E(\text{sMAX})$$

Fig. 1 The expression of MSSC

Phone Codes. Therefore, when generating an MSSC from a word or phrase, it is necessary to apply some rules in which the mouth shapes change.

3.2 Generation Rules of Mouth Shapes Sequence Code

In order to generate a correct MSSC for a word or phrase with more than two phones, simply connect the Phone Codes corresponding to the phone of the word or the phrase from Table 2 and 3 as preparations and generate an MSSC. At that time, the MS Codes "$-\star$" should be temporarily given because the mouth shape can not given to the geminate consonant nor the syllabic nasal, and replace it with the MS Code at a stage that mouth shape is fixed. In an MSSC, because a combination of odd-numbered and even-numbered MS Codes turns out to be a mouth shape segment, we call these combinations as the first mouth shape segment, the second mouth shape segment, the third mouth shape segment, $\cdots$, the n-th mouth shape segment sequentially from the top.

Next, we define an expression to show an MS Code in an MSSC. If $s = 1,2,3,\cdots,\text{sMAX}$ (sMAX; number of the mouth shape segments of a word or phrase), we assume that $c_B(s)$ shows an MS Code for BeMS of the s-th mouth shape segment and $c_E(s)$ shows an MS Code for EMS of the s-th mouth shape segment. In the previous example of "asahi" (number of the mouth shape segments; 3), $c_B(s)$ and $c_E(s)$ of the expression with $s = 1,2,3$ will be shown in Table 4. The description of general MSSC is shown in Figure 1.

Table 4 MS Code of each mouth shape segment for "asahi"

s	$c_B(s)$	$c_E(s)$
1	-	A
2	I	A
3	-	I

As for the mouth shape change by the combination of consecutive phones, we make those rules logical by using the expression and MS Code suggested in the preceding section.

Mouth shape change rule 1

If $c_E(s) = c_E(s-1)$ and $c_B(s) =$ '-' then $c_B(s)$ and $c_E(s)$ are deleted.

For example, the MSSC simply connected becomes "-A-A-I" when uttering the Japanese word "akari" (a light in English). Here, the expression, $c_E(2) = c_E(1)$ becomes true because $c_E(2) =$ 'A' and $c_E(1) =$ 'A'. Moreover, $c_B(2) =$ '-' becomes true. Therefore, $c_B(2)$ and $c_E(2)$ will be deleted. As a result, the MSSC of "akari" becomes "-A-I".

In this case, "akari" is made of a three moras word, but is made of two mouth shape segments. In this way, number of the mouth shape segments and number of moras do not match each other on the cases that Simple Mouth Shape Phone from the same column continues.

Mouth shape change rule 2
If $c_B(s) = c_E(s-1)$ then $c_B(s) = $ '-'.

For example, the MSSC simply connected becomes "-IIE" when uttering the Japanese word "ise" (a place name). Here, $c_B(2) = c_E(1)$ becomes true because $c_B(2) = $ 'I' and $c_E(1) = $ 'I'. Therefore, $c_B(2) = $ '-'. As a result, the MSSC of "ise" becomes "-I-E".

Mouth shape change rule 3
If $c_E(s) = $ '*' and $c_B(s+1) = $ 'X' then $c_E(s) = $ 'X' and $c_B(s+1) = $ '-'.

For example, the MSSC simply connected becomes "-O-*XU" when uttering the Japanese word "koppu" (glass). Here, $c_E(2) = $ '*' and $c_B(3) = $ 'X' become true. Therefore, $c_E(2) = $ 'X' and $c_B(3) = $ '-'. As a result, the MSSC of "koppu" becomes "-O-X-U".

Mouth shape change rule 4
If $c_E(s) = $ '*' and $c_E(s-1) = $ 'A' or $c_E(s) = $ '*' and $c_E(s-1) = $ 'E' then $c_E(s) = $ 'I'.

For example, the MSSC simply connected becomes "-E-*UO" when uttering "endo" (end). Here, $c_E(2) = $ '*' and $c_E(1) = $ 'E' become true. Therefore, $c_E(2) = $ 'I'. As a result, the MSSC of "endo" becomes "-E-IUO".

Mouth shape change rule 5
If $c_E(s) = $ '*' and $c_E(s-1) = $ 'O' then $c_E(s) = $ 'U'.

For example, the MSSC simply connected becomes "UO-*-I" when uttering the Japanese word "tokki" (a projection). Here, $c_E(2) = $ '*' and $c_E(1) = $ 'O' become true. Therefore, $c_E(2) = $ 'U'. As a result, the MSSC of "tokki" becomes "UO-U-I".

Mouth shape change rule 6
If $c_E(s) = $ '*' and $c_E(s-1) = $ 'I' or $c_E(s) = $ '*' and $c_E(s-1) = $ 'U' then $c_B(s)$ and $c_E(s)$ are deleted.

For example, the MSSC simply connected becomes "-I-*UO" when uttering the Japanese word "kinjo" (neighborhood). Here, $c_E(2) = $ '*' and $c_E(1) = $ 'I' become true. Therefore, $c_B(2)$ and $c_E(2)$ are deleted. As a result, the MSSC of "kinjo" becomes "-IUO".

But, the mouth shape change rule 3 will be applied prior to the mouth shape change rule 4 to 6. And we can generate a correct MSSC from a word or phrase by using the mouth shape change rule 1 to 6, Table 2 and 3.

62 T. Miyazaki, T. Nakashima, and N. Ishii

4 Experiments

Test words used in this experiment, those Japanese syllabaries and number of moras
for them are shown in Table 5. The general flow diagram of the program that
the mouth shape change rules are incorporated to generate an MSSC is shown in
Figure 2.

Table 5 Test words

#	word (in English)	Japanese syllabary	moras
1	camera	kamera	3
2	bicycle	zitensya	4
3	the Beijing Olympics	pekinorinpikku	9
4	desktop PC	desukutoppupasokon	10

4.1 Experimantal Result

The MSSC generated by "kamera" became "-AXEIA". It was confirmed that right
MSSC was generated from real utterance. Because the simply connected MSSC
which was generated before applying mouth shape change rules "-AXEIA", it was
confirmed that any mouth shape change rules were not applied to this word.

About "zitensya", some of mouth shape change rules were applied. The application process of the mouth shape change rules is shown in Table 6. The numbers of
the mouth shape change rule applied to each column of "applied rule" in this table
and the MSSC after applying a mouth shape change rule to each column of "(tem-
porally) MSSC" are shown. In addition, the MSSC shown in row #0 is the MSSC
simply connected as previous preparations.

Table 6 The process of applying the mouth shape change rules to the test word #2

#	applied rule	(temporally) MSSC
0	–	-IIE-*IA
1	2	-I-E-*IA
2	4	-I-E-IIA
3	2	-I-E-I-A

As a result, the mouth shape change rule 2 was applied at first to the second mouth
shape segment. Next, the mouth shape change rule 4 was successively applied to the
third mouth shape segment. Finally, the mouth shape change rule 2 was applied to
the fourth mouth shape segment. As a result, an MSSC "-I-E-I-A" is generated.
By comparing with actual word utterance, we really confirmed that the generated
MSSC expressed right mouth shapes order.

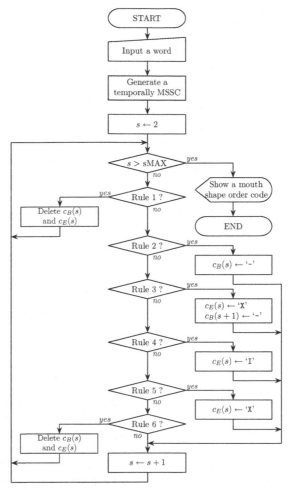

"Rule n ?"; Is the mouth change rule n applicable?

Fig. 2 Flow diagram to generate an MSSC

In the same manner, the application process of the mouth shape change rules about the word #3 is shown in Table 7. As a result, the mouth shape change rule 6 was applied to the third mouth shape segment first. Next, the mouth shape change rule 3 was applied to the fifth mouth shape segment. It was also understood here that the mouth shape change rule 3 was applied although the mouth shape change rule 3 and 6 were objective of the applications in the fifth mouth shape segment. Finally, the mouth shape change rule 6 was applied to the seventh mouth shape segment. As a result, an MSSC "XE-I-O-I-X-I-U" was generated. By uttering actual word, we confirmed that the generated MSSC expressed right mouth shape order.

The application process of the mouth shape change rules about the word #4 is shown in Table 8. In this word, the mouth shape change rule 1, 2, 3 and 5 were

Table 7 The process of applying the mouth shape change rules to the test word #3

#	applied rule (temporally) MSSC	
0	–	XE-I-*-O-I-*XI-*-U
1	6	XE-I-O-I-*XI-*-U
2	3	XE-I-O-I-X-I-*-U
3	6	XE-I-O-I-X-I-U

Table 8 The process of applying the mouth shape change rules to the test word #4

#	applied rule (temporally) MSSC	
0	–	IE-U-UUO-*XUXAUO-O-*
1	1	IE-UUO-*XUXAUO-O-*
2	2	IE-U-O-*XUXAUO-O-*
3	3	IE-U-O-X-UXAUO-O-*
4	1	IE-U-O-X-UXAUO-*
5	5	IE-U-O-X-UXAUO-U

applied, and the generated MSSC was "IE-U-O-X-UXAUO-U". And, from actual utterance, we confirmed that it was right MSSC.

5   Conclusion

In this study, we defined the distinctive mouth shapes at which lip-reading skill holders stare as Basic Mouth Shape. In addition, by classifying the mouth shapes at an utterance into the Beginning Mouth Shape and the End Mouth Shape, we expressed combination patterns of the two kinds of mouths shape for all the Japanese phones. And we defined the Mouth Shape Code for the BaMS and suggested the Mouth Shapes Sequence Code which expressed BeMS and EMS sequentially formed at the time of a word or phrase utterance. We have been able to generate an MSSC form a word or phrase without actual utterance by using the patterns of BeMS and EMS of Japanese phones. However, we defined six mouth shape change rules with the expressions for the patterns of the BeMS and the EMS because the mouth shape could change when uttering consecutive phones of a word. Applying these rules have made it possible to generate right MSSC even from a word or phrase.

In this way, we suggested the coding method of the distinctive mouth shapes for machine lip-reading and the indication method of the mouth shape change information based on the knowledge of the lip-reading skill holders here, but there are some issues to solve. First, it is remained to discuss the examination about the point where the method of utterance of the real word or phrase will vary. For example, there are two cases; 1) uttering a phone by dividing, 2) uttering a phone in succession.

Second, it is also remained to discuss about the point where some MSSC generated by the different word or phrase may result in the same codes. Two such cases could exist. 1) The mouth shape change rule 1 is applied to some continuing mouth shape segments at a state where MSSC are generated. For example, as for the Japanese word "ikigire" (gasp in English) and "ie" (house), both of the MSSC become "-I-E". It is possible to distinguish these from the MSSC, but their identification is possible because their duration time of the mouth shape /i/ formed is different each other at the actual utterance. Second example; The case that the duration of the phones such as the Japanese words "tabako" (cigarette) and "tamago" (egg) is identical (the MSSC for them is "IAXA-O"). It is difficult to distinguish the allophone words uttered by same mouth shapes even if information about the length of the mouth shape is used for these words. For those words, it is necessary to examine methods to infer candidate words from a theme and/or the content of the story.

References

1. Kenji, M., Pentland, A.: Automatic Lip-reading by Optical-flow Analysis. The Transactions of the Institute of Electronics, Information and Communication Engineers J73-D-II(6), 796–803 (1990) (in Japanese)
2. Takashi, O., Teruhiko, O.: Automatic Lipreading of Station Names Using Optical Flow and HMM. Technical Report of IEICE PRMU 102(471), 25–30 (2002) (in Japanese)
3. Mang, L., Issei, Y., Yoshihiro, K., Hidemitsu, O.: Automatic Lipreading by Subspace Method. Technical Report of IEICE PRMU 97(251), 9–14 (1997) (in Japanese)
4. Takeshi, S., Ryosuke, K.: Lip Reading Based on Trajectory Feature. The IEICE Transactions on Information and Systems (Japanese edition) J90-D(4), 1105–1114 (2007) (in Japanese)
5. Kimiyasu, K., Keiichi, U.: An Utered Word Recognition Using Lip Image Information. The Transactions of the Institute of Electronics, Information and Communication Engineers J76-D-II(3), 812–814 (1993) (in Japanese)
6. Akihiro, O., Yoshitaka, H., Kenji, O., Toshihiko, M.: Speech Recognition Based on Integration of Visual and Auditory Information. Transactions of Information Processing Society of Japan 39(12), 3232–3241 (1998) (in Japanese)
7. Yasuyuki, N., Moritoshi, A.: Lipreading Method Using Color Extraction Method and Eigenspace Technique. The Transactions of the Institute of Electronics, Information and Communication Engineers J85-D-II(12), 1813–1822 (2002) (in Japanese)

Key Predistribution Scheme Using Finite Fields and Reed Muller Codes

Pinaki Sarkar and Morshed U. Chowdhury

Abstract. Resource constraint sensors of a *Wireless Sensor Network (WSN)* cannot afford the use of costly encryption techniques like *public key* while dealing with *sensitive data*. So *symmetric key encryption* techniques are preferred where it is essential to have the same cryptographic key between communicating parties. To this end, keys are preloaded into the nodes before deployment and are to be established once they get deployed in the *target area*. This entire process is called key predistribution. In this paper we propose one such scheme using *unique factorization of polynomials over Finite Fields*. To the best of our knowledge such an elegant use of Algebra is being done for the first time in WSN literature. The best part of the scheme is large number of node support with very small and uniform *key ring* per node. However the resiliency is not good. For this reason we use a special technique based on Reed Muller codes proposed recently by Sarkar, Saha and Chowdhury in 2010. The *combined scheme* has good resiliency with huge node support using very less keys per node.

1 Introduction

Wireless sensor networks consist of tiny sensor nodes that have very limited battery power, less amount of storage, low computational power and they are scattered in large numbers over a vast region. The sensors communicate between each other and with the base station via radio frequencies. These networks are used in scientific purposes like smoke detection, wild fire detection, seismic activity monitoring, etc.

Pinaki Sarkar
Department of Mathematics, Jadavpur University, Kolkata, India
e-mail: pinakisark@gmail.com

Morshed U. Chowdhury
School of Information Technology,
Deakin University-Melbourne Campus, Melbourne, Australia
e-mail: morshed.chowdhury@deakin.edu.au

R. Lee (Ed.): Software Eng., Artificial Intelligence, NPD 2011, SCI 368, pp. 67–79.
springerlink.com © Springer-Verlag Berlin Heidelberg 2011

Besides they have large application in military purposes, for instance monitoring enemy movements. Clearly, the nodes deal with very sensitive data and can communicate within a special range called Radio Frequency range. Since sensors are deployed unattended over the target area this makes them physically insecure and prone to adversarial attacks. Thus arises the need of secure communication model to circumvent these attacks.

A secure communication model makes use of (low cost) cryptographic primitives. Existing schemes like Kerberos [15] and public key cryptography [8] are not suitable to this kind of resource constrained system because of the inherent cost associated to these schemes.

Eschenauer and Gligor in [7] were first to suggest that key predistribution can be a effective way to establish keys securely with low cost. Key predistribution is a symmetric key approach where two communicating nodes share a common secret key. The messages are encrypted and decrypted using the same secret key. As online key exchange is rather costly, both sender and receiver nodes must be preloaded with the same key. Thus prior to actual exchange of message, steps involved in key predistribution are:

- Before deployment a centralized authority called Base Station or Key Distribution Server (KDS) preloads keys into individual sensors to form their *key rings* or *key chain* from a collection of keys called the *key pool*.
- Establishing shared keys between nodes before actually communication of nodes (aka shared key discovery).
- Establishing a path-key between two nodes that do not have a shared key.

1.1 Related Works

Key predistribution schemes are of two types viz. Probabilistic and deterministic. In probabilistic key predistribution scheme keys are selected at random from a large key-pool and loaded into sensor nodes. In deterministic key predistribution scheme as the name implies, a deterministic pattern is used to achieve the goal. Xu et al. [17] showed that probabilistic schemes perform marginally better than deterministic schemes.

The above mentioned work of Eschenauer and Gligor [7] provides a probabilistic scheme. An excellent survey of such probabilistic schemes can be found in a report published in 2005 by Çampte and Yener [4].

On the other hand deterministic key predistribution schemes make key establishment simpler while ensuring better connectivity than probabilistic schemes. This fact was established by Lee and Stinson [10, 11]. Another deterministic schemes can be found in [2, 12]. Such schemes doesn't always have full connectivity. For instance the first two doesn't while the remaining one has full connectivity. Chakraborty et al. [5] suggested a hybrid scheme where blocks obtained using transversal designs proposed in [10, 11] were randomly merged to form new nodes. Clearly this increases the *key rings* of each node while improving on resiliency and communication probability.

A code based key management system was proposed by Al-Shurman and Yoo [14] where matrices have been used along with a random vector to generate codeword which is the secret key chain. With certain probabilities the design satisfies Cover-free-family (CFF). However neither the problem of connectivity in a network nor is the scenario of node compromise is addressed.

Later in 2009, an elegant key predistribution scheme using Reed Solomon codes was proposed by Ruj and Roy [13]. Though the scheme did not provide full connectivity amongst nodes, however its resiliency and scalability is worth appreciating. The beauty of the codes is the use of polynomials during key establishment phase. This prompted us to think in a similar direction.

1.2   Our Contributions

In this paper a key predistribution scheme is being proposed. The *key pool* is taken to be the element of the finite field $\mathbb{Z}_p$ or $\mathbb{F}_p$. Hence the *key ring* of each node is a subset of size k of $\mathbb{F}_p$. Now using the unique factorization property of polynomials rings over $\mathbb{F}_p$ one can construct the node identifiers (ids). To ensure full connectivity one must choose $k = \lceil p/2 \rceil, k \neq 2$. Obviously, this hampers the resiliency immensely. However one can now apply a novel trick introduced in the literature by Sarkar et al. [16] in 2010. The combined idea ensures a large network having decent resiliency with very less number of keys per node.

One observes that the in the key predistribution scheme, the key pool can be extended to be elements of any finite field $\mathbb{F}_q$ where $q = p^r$ for some $r \geq 1$ and p : prime. For $q = 2^r$, i.e., $\mathbb{F}_{2^r}$ take $k = \lceil p/2 \rceil + 1$.

1.3   Notations

Throughout the paper we shall use the term "Uncompromised nodes" to means nodes that are not compromised. The words "communication" and "connectivity/connection" are sometimes abbreviated to com. and con. respectively. The terms "communication model/scheme" and "key predistribution model/scheme" will mean the same.

Before explicitly explaining the scheme, some basic notions like communication, connectivity, respective keys and communication radius (aka *Radio Frequency range*) are required. One can refer to [16, section II] for these concepts.

2   Set System: Balanced Block Design

This section is devoted to describe the key predistribution problem based on generic Balanced Block Designs.

Set system or design [11] is a pair $(\mathbb{X}, \mathbb{A})$ where $\mathbb{A}$ is a set of subsets of $\mathbb{X}$ called blocks. The elements of $\mathbb{X}$ are called varieties. Balanced Incomplete Block Designs *(BIBD)* (v, b, r, k, λ) are designs satisfying the following conditions:

- $|\mathbb{X}| = v$, $|\mathbb{A}| = b$.
- Each subset in $\mathbb{A}$ contains exactly k elements (*rank*).
- Each variety in $\mathbb{X}$ occurs in r many blocks (*degree*).
- Each pair of varieties in $\mathbb{X}$ is contained in exactly λ blocks of $\mathbb{A}$.

Thus $(\mathbb{X}, \mathbb{A})$ is a set system with $(\mathbb{X} = \{x_i : 1 \le i \le v\}$ and $\mathbb{A} = \{A_j : 1 \le j \le b\}$. *Dual set system* of $(\mathbb{X}, \mathbb{A})$ is any set system isomorphic to the set system $(\mathbb{X}, \mathbb{A})$ with:

- $\mathbb{X} = \{x_j : 1 \le j \le b\}$
- $\mathbb{A} = \{A_i : 1 \le i \le v\}$ where
- $x_j \in A_i$ for original *BIBD* $\Longleftrightarrow x_i \in A_j$ for its dual.

Thus it follows that if one considers the dual of a BIBD (v, b, r, k, λ), one arrives at a design containing b varieties and v blocks. Each block of the dual BIBD contains exactly r varieties and each variety occurring in exactly k blocks. Clearly any two blocks contain λ elements in common. When $v = b$, the BIBD is called a symmetric BIBD and is denoted by $SBIBD(v, k; \lambda)$.

In case $b = \binom{v}{k}$, i.e. one considers all possible blocks of a set system, one arrives at a Balanced Block Design (*BBD*). One such design will give our key predistribution scheme.

3 Key Predistribution Scheme

Description of a key predistribution scheme based on unique factorization of polynomials over finite fields will be presented here. One can refer to any standard text book of Undergraduate Algebra like [9, 1] for detail about Unique Factorization Domains (*UFD*).

3.1 Distribution of Keys: Key Pool and Key Rings

Consider the finite field $\mathbb{Z}_p$ or better $\mathbb{F}_p$ where $p \ne 2$ is a prime. One usually denotes the elements of the set $\mathbb{Z}_p$ as $0, 1, 2, \ldots, p-1$. These will be treated as the *key pool*.

Subsets of size k are picked up randomly from the above set of p elements. These are mapped to the nodes as key identifiers corresponding to k keys in their respective *key rings*. Thus a total of $\binom{p}{k}$ nodes can be supported with k keys per node.. Considering k is very small compared to $\binom{p}{k}$ (k close to $p/2$), one gets huge node support with less keys per node.

Direct communication amongst any given pair of nodes is essential in ensuring fast and inexpensive communication. While choosing the number of keys per node, i.e., exact value of k, the above fact has to be kept in mind. *Pigeon Hole principle* assures with $k = \lceil p/2 \rceil$, any two *key rings* will have non empty intersection. Formally stating:

Theorem 1. *Any two subsets of $\mathbb{F}_p$ of cardinality $k = \lceil p/2 \rceil$ has non-empty intersection where $p \ne 2$.*

Proof. The proof will be done by contradiction using *pigeon hole principle*. Assume two arbitrary subsets of $\mathbb{Z}_p$, say A & B of size $k = \lceil p/2 \rceil$ has empty intersection, i.e., $A \cap B = \emptyset$. Invoking principle of inclusion/exclusion one has $|A \cup B| = |A| + |B| - |A \cap B|$. Since $|A| = |B| = k = \lceil p/2 \rceil$ (by choice) and $|A \cap B|$ (assumed), one concludes: $|A \cup B| = \lceil p/2 \rceil + \lceil p/2 \rceil - 0 = \frac{p+1}{2} + \frac{p+1}{2} = p + 1 > p$.

However since A and B are subsets of $\mathbb{Z}_p$, $|A \cup B|$ should also be a subset of $\mathbb{Z}_p$. Thus $|A \cup B|$ cannot have more than p distinct elements contradicting the above calculation of $|A \cup B| = p + 1$.

In terms of *pigeon hole principle* there are $2k$ pigeon & p holes. So for $k \geq \lceil p/2 \rceil$, two pigeons share the same hole. This proof easily generalizes to $\mathbb{F}_q$

Corollary 1. Choosing $k = \lceil p/2 \rceil$ assures that any two *key rings* will have at least one common key where *key pool* is $\mathbb{Z}_p$.

Proof. The proof readily follows from the above theorem by observing that one can treat the two *key rings* as the arbitrary sets A and B of the above theorem.

Remark 1

- To avoid greater collision as well as reduce memory usage the minimum number: $k = \lceil p/2 \rceil$ is chosen to be *key rings*.
- The key predistribution scheme can be extended to any finite field $\mathbb{F}_q$ where $q = p^r$ for some $r \geq 1$ and p : prime. The key pool can be the elements of the finite field and the *key ring* are any subset of size $k = \lceil q/2 \rceil$ Hence total number of nodes $= \binom{q}{\lceil q/2 \rceil}$.
- For $q = 2^r$; $r \geq 1$ key ring size $= k = \lceil q/2 \rceil + 1$

3.2 Node Identifiers: Key Establishment: U.F.D.

In any deterministic key predistribution, each node is assigned with a unique node identifier (node ids.) These node ids are later utilized during key establishment phase to find out the common keys.

In the present scheme consider two arbitrary distinct nodes A and B having key rings $\{\alpha_1, \alpha_2 \ldots \alpha_k\}$ and $\{\beta_1, \beta_2, \ldots, \beta_k\}$ where $k = \lceil p/2 \rceil$. Consider the polynomials $\alpha = \prod_{i=1}^{k}(x - \alpha_i)$ and $\beta = \prod_{i=1}^{k}(x - \beta_i)$ over $\mathbb{Z}_p$. That is take the product of all linear factors and go modulo p. Since $\mathbb{Z}_p[x]$ is a *unique factorization domain* [1, 9] and key rings any two differs by at least one key, hence key id, it readily follows that $\alpha \neq \beta$. For any two it is assured that the polynomial thus formed by multiplying all the linear factors $(x-$ *key ids*) & going modulo p is unique. It should be noted that each polynomial can be expressed as a $k = \lceil p/2 \rceil$ tuple. Of course these are all k^{th} degree *monic* polynomials. So though they have $k + 1$ terms, the leading term is always 1. The idea is to treat these k tuples as node ids of each node.

Clearly after deployment in the target area when the key establishment has to be done, these node ids (k tuples) of each node can be passed onto the other nodes. On receiving each other's ids the nodes form the polynomials whose coefficients are

the received k-tuples. Substituting the key ids into the polynomials will enable each
node to find out the common roots, hence common key id and thus keys. Here it is
important to observe that the roots of the polynomials are all *linear* & *distinct* as is
clear from construction.

One readily sees that the scheme described above can be easily extended to any
finite field. Meaning take $\mathbb{F}_q$ in place of $\mathbb{F}_p$ where $q = p^r, r \geq 1, p : prime \neq 2$ For
$\mathbb{F}_{2^r}$ taking the key rings to be of size $k = 2^{r-1} + 1$ ensures non-empty intersection
by pigeon-hole principle. Otherwise the construction is absolutely same.

4 Weakness of the Scheme

There are some obvious weakness of this key predistribution strategy like in most
deterministic schemes involving node ids. Firstly the node ids are send through in
secure channel. Key establishment is generally a very fast process. So capturing
nodes may be difficult this stage. However attacking the channel is still feasible. As
the node ids are sent unencrypted and the adversary is assumed to know the system,
she can do the following:

- get the node ids, i.e., co-efficients of the node's polynomial from open channel.
- compute the polynomials.
- substitute all elements of the finite field $\mathbb{F}_q$ or may be even equate the polynomials to find out the key ids.
- thus she knows the exact key sharing strategy between nodes. Selective node
  attack is thus possible as described in [13].

Other than this there is a inherent problem with the current scheme. Here the key
pool is of size p and each key ring of size $k = \lceil p/2 \rceil$. Thus with decent probability
capturing two nodes can yield the the entire key pool.

However one can use a clever trick recent proposed by Sarkar, Saha and Chowdhury [16] to increase the resiliency of the system.

5 Combined Communication Scheme

A combination of the above key predistribution scheme with the black box design
suggested by Sarkar et al. in [16] after appropriate modifications is presented here.
The resultant combined communication model has good resiliency and can support
lot of node with small key rings for a WSN.

Before proceeding further some practical computation in terms of key ring size
and node support needs to be done.

5.1 Some Practical Considerations

From Table 1, it is clearly visible that for very small key rings the node support is
tremendous. The '*' marks indicate that these are the special cases of powers of 2

Table 1 Key pool, Key ring VS. Node support for prime power ≤ 25

Choice of prime power (key pool size) $= q = p^r$	Size of key ring $= k = \lceil q/2 \rceil$	Node support $= \binom{q}{k}$
7	4	35
$9 = 3^2$	5	126
11	6	462
13	7	1716
$16 = 2^4$	9*	11440
17	9	24310
19	10	92378
23	12	1352078
$25 = 5^2$	13	5200300

where $k = 2^{r-1} + 1, r = 4$. These practical calculations will play a very important role in designing the ultimate combined model using technique of [16]. By definition

$$\binom{q}{\lceil q/2 \rceil} = \frac{q!}{\frac{q-1}{2}! \frac{q+1}{2}!}$$

which can be expressed differently as is demonstrated later.

5.2 Connectivity Model

While trying to apply the novel concept of treating connectivity and communication separately and then giving adequate security to the connectivity model, one has to fix up the cluster sizes. This depends on the storage, computation and other capabilities of the Cluster Heads (CHs).

In the work presented in [16], Sarkar, Saha and Chowdhury have cunningly utilized the construction of Reed Muller codes. Such codes are well described in Cooke's work [6]. Necessary changes can be found in [16, section IV]. The connectivity matrices for a cluster with one CH and m nodes were constructed by looking at $\mathbb{Z}_2[X_1, X_2, \ldots, X_m]$. This polynomial ring was treated as a vector space over $\mathbb{Z}_2$. The bit pattern $2^i 1's, 2^i 0s, \ldots, 2^i 0's$ (length=2^m) of basic monomials x_i were assigned to the nodes i $1 \leq i \leq m$. Whereas the CH (can be Base Station) received the vector $2^m 1's$. Such matrices suits upper tier of the hierarchical structure. However provision had been made for weaker nodes at the last level. For this, one needs to combine polynomials as per connectivity demand. Rest of the details can be found in [16, section IV] and is omitted here as the same needs to be followed. The major job now is to see how their inspiring work can help in the present case.

Viewing differently for $q \neq 2$, one can split,

$$\binom{q}{\lceil q/2 \rceil} = \frac{q!}{\frac{q-1}{2}! \frac{q+1}{2}!} =$$

$$\{\frac{q(q-1)}{\frac{q-1}{2}\frac{q-3}{2}}\}\frac{(q-2)(q-3)\ldots[(q+1)/2]}{\frac{q-5}{2}!}$$

one can easily assume that each CH can accommodate roughly $m = \dfrac{q(q-1)}{\frac{q-1}{2}\frac{q-3}{2}}$ nodes.

The connectivity matrix is assigned above m number of variables in the present case.

Clearly one can choose more number of nodes per cluster in order to decrease the number of cluster heads. Thus obtaining different value of m. One can even select unequal nodes per cluster.

Remark 2. Providing security to additional number

$$O(\frac{(q-2)(q-3)\ldots[(q+1)/2]}{\frac{q-5}{2}!})$$

of CHs seems quite practical. The height of tree is a logarithmic function of q or better still number of nodes, so message exchange speed is quite acceptable.

Connectivity Network Structure

Figure 1 gives a sketch of the connectivity structure of proposed scheme in accordance with Sarkar et al. model [16]. Clearly this model is heterogeneous in the sense

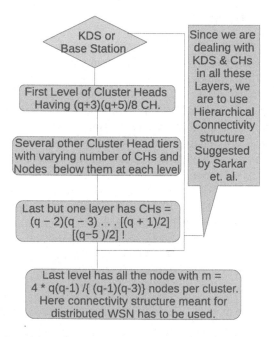

Fig. 1 Connectivity Network Structure for the Key Predistribution Scheme proposed in this paper using the model proposed by Sarkar, Saha and Chowdhury in [16].

that number of sensors or CHs varies from cluster to cluster. As stated in their work different cryptosystem can used at different tiers.

6 Message Sending Protocol

For sending a message from node N_i to node N_j for some fixed $1 \leq i \neq j \leq \binom{q}{k}$ where $k = \lceil q/2 \rceil$, q : a prime power, the following protocol is to be executed:

Fix up one common communication key between N_i and N_j. Call it μ_{ij}.

N_i encrypts the message with this key $\mu_{i,j}$.

if N_i and N_j share a connectivity key **then**

> The message encrypted with com. key is again encrypted with the shared con. key and send directly to node N_j.
>
> N_j decrypts the outer encryption done using the con. key common to both the nodes.

else

> node N_i uses the con. key that it shares with its Cluster Head and send the doubly encrypted message to its CH.
>
> **if** node N_j lies in the same cluster **then**
>
> > After decrypting with N_i's con. key and encrypting with N_j's con. key, the common CH directly send it to node N_j .
> >
> > N_j decrypts outer encryption done using the con. key that it shares with the (common) CH.
>
> **else**
>
> > The doubly encrypted message from N_i is decrypted using N_i's con. key at the CH of N_i.
> >
> > It is re-encrypted at CH of N_i using the con. key shared with Cluster Head of N_j.
> >
> > Send doubly encrypted message to CH of N_j.
> >
> > Cluster Head of N_j then decrypts it with the con. key shared with the cluster head of N_i.
> >
> > CH of N_j encrypts it using the shared con. key with N_j.
> >
> > Send the doubly encrypted message to N_j.
> >
> > N_j will first decrypt the outer encryption done using the con. key of its CH (not N_i's).
> >
> > N_j decrypts outer encryption done using the con. key common to both the nodes.
> >
> > N_j decrypts outer encryption done using the con. key common to both the nodes.
>
> **end if**

end if

Finally N_j uses the chosen common com. key μ_{ij} shared with N_i to decrypt and read the message.

---

7   Network Parameters

Theoretical analysis of some important aspects of the combined communication model will be present here. Main topics of interests include *communication probability, computational overhead, resiliency and scalability.*

7.1   Communication Probability and Overhead

Communication Probability is defined to be the probability that any given pair of nodes can communicate directly. From corollary to **Theorem 1**, it is obvious that *Communication Probability is* 1.

Communication overhead measures the computational complexity of first the key establishment and then message exchange protocols. During *key establishment* only computation done is substitution of key ids which are elements of $\mathbb{F}_q$ in the given polynomial and verifying if the computed value is 0. The complexity of the entire process primarily reduces to computing $\lceil p^r/2 \rceil$ power of elements of $\mathbb{F}_{p^r}$. One can invoke Horner's algorithm or still better Bernstein's improvement to reduces the complexity of algorithm involved. Such algorithms are well explained in Bernstein's work presented in [3].

Applying Horner's rule would require roughly $\lceil \frac{p^r}{2} \rceil$ multiplication & $\lceil \frac{p^r}{2} \rceil$ addition. Applying Horner's algorithm, key establishment will take $O(\lceil \frac{p^r}{2} \rceil)$ steps.

Whereas Bernstein reduced the number of operations of this polynomial evaluation to $\lceil p^r/4 \rceil$ multiplication, $\lceil p^r/2 \rceil$ addition and $ln\lceil p^r/2 \rceil$ squaring. Moreover the squaring can be done in parallel to the multiplications. The key establishment can be thus accomplished in $O(\lceil p^r/4 \rceil)$ steps.

As for message exchange, at node level double encryption and decryption is done. So complexity is just twice than the usual techniques which employ single encryption. Of course same cryptosystem (AES-128 bit) can be utilized. Only the keys need to be different.

7.2   Resilience and Scalability

Like in the work of Sarkar et al. [16] the resilience of the system is heavily dependent on the CHs. In a similar logic to their proposed model, the system is totally resilient against only ordinary node capture or only CH capture.

However the situation can become very bad in case of capture of both nodes and CHs. When a CH is captured the massages passing through it are encrypted with only communication keys. If any of these keys are exposed by virtue of some node capture anywhere in the network, message exchange of the nodes below the captured CHs using these keys becomes unsafe. It is also easy to see that with probability = $\frac{\binom{q}{k}k}{\binom{q}{k}^2} = \frac{k}{\binom{q}{k}}$ ($k = \lceil q/2 \rceil$) the entire key pool is exposed when two nodes gets captured. Thus quite evidently the security of the combined model is based on the security of CHs. Since CHs are much less in number, one can provide extra security to them against physical capture.

As for scalability, the initial key predistribution scheme is not scalable by the look of things. However one can look at it from a different angle. Since there are too many nodes with too few keys per node, one may need much less number of node support. As such, one has to choose the nodes cleverly having key rings intersecting at as less number of keys as possible. Of course any given pair of key rings will intersect by the construction. This not only ensures good scalability but also assures better resiliency to start with.

8 Conclusion

It is challenging problem in the literature of Wireless Sensor Networks (WSN) to give a key predistribution scheme which has good resiliency and is equally scalability. Though the initial scheme presented in this paper is not that good in terms of resiliency, it supports huge number of nodes with very small ($k = \lceil p/2 \rceil$) and uniform sized *key ring* per node 3. The elegant use of simple Algebraic tools like *unique factorization* in polynomial rings over Finite Fields and *pigeon hole principle* is worth noticing.

Later when the proposed key predistribution scheme is combined with the black box design of Sarkar et al. [16], the resiliency is much improved. This paves a direction which may fulfill the target of getting above mentioned scheme.

9 Future Work

Continuing from the discussions of the previous section one can readily visualize some immediate future research directions. The size of key rings as compared to the size of the key is quite large. This results in weakening the resiliency of the proposed key predistribution scheme. To guard against this weakness Sarkar et al. [16] trick was invoked. Clearly the security of the system has primarily shifted to guard the CHs from physical capture. However even in an unlikely event of physical CH capture, some node has to be captured anywhere in the network. Only then the nodes under that CH gets affected. To this end, its important to have a key predistribution scheme where inherent resiliency is much better.

The scalability issue should be kept in mind. It should be highlighted that the present scheme provide a huge node support with very less keys per node. One can easily compromise on this. Prior to this proposal most schemes which supported N nodes required $O(\sqrt{N})$ keys per node. So even if one designs a resilient and scalable scheme with N nodes having $O(\sqrt[3]{N})$ keys per node, a good job is done. In this regard one can use the idea of this scheme as a stepping stone. One may look at some special subsets of the entire node space $\mathbb{F}_p$, having much less elements. Then concentrate on building up key rings intersecting with each other but at less number of places using Mathematical technique. This will assure much resilient scheme with full connectivity. The scalability will follow by allowing the new nodes to come in with some extra keys.

Acknowledgement

We want to express our gratitude to University Grants Commission (UGC), India for financially supporting the doctoral program of Mr. Pinaki Sarkar. This work is meant to be a part of the doctoral thesis of Mr. Pinaki Sarkar.

Special word of appreciation is due to Miss. Amrita Saha of IIT, Bombay for actively discussing various aspects of this publication including suggestion of use of the novel trick introduced in [16] where she contributed significantly in developing Mr. Pinaki Sarkar's original idea under Dr. Morshed U Choudhury's guidance.

References

1. Artin, M.: Algebra. Prentice Hall, India (2010) ISBN-81-203-0871-9
2. Bag, S., Ruj, S.: Key Distribution in Wireless Sensor Networks using Finite Affine Plane. In: AINA 2011, pp. 436–442. IEEE Computer Society, Los Alamitos (2011)
3. Bernstein, D.: Polynomial evaluation and message authentication (2007), http://cr.yp.to/antiforgery/pema-20071022.pdf
4. Çamtepe, S.A., Yener, B.: Key distribution mechanisms for wireless sensor networks: A survey. Technical Report (TR-05-07), Rensselaer Polytechnic Institute, Computer Science Department (2005)
5. Chakrabarti, D., Maitra, S., Roy, B.: A key pre-distribution scheme for wireless sensor networks: merging blocks in combinatorial design. International Journal of Information Security 5(2), 105–114 (2006)
6. Cooke, B.: Reed Muller Error Correcting Codes. MIT Undergraduate J. of Mathematics (1999)
7. Eschenauer, L., Gligor, V.D.: A key-management scheme for distributed sensor networks. In: ACM Conference on Computer and Communications Security, pp. 41–47 (2002)
8. Gura, N., Patel, A., Wander, A., Eberle, H., Shantz, S.C.: Comparing elliptic curve cryptography and RSA on 8-bit cPUs. In: Joye, M., Quisquater, J.-J. (eds.) CHES 2004. LNCS, vol. 3156, pp. 119–132. Springer, Heidelberg (2004)
9. Gallian, J.: Contemporary Abstract Algebra. Narosa Publishing House, India (2004) ISBN-81-7319-269-3
10. Lee, J.-Y., Stinson, D.R.: Deterministic key predistribution schemes for distributed sensor networks. In: Handschuh, H., Hasan, M.A. (eds.) SAC 2004. LNCS, vol. 3357, pp. 294–307. Springer, Heidelberg (2004)
11. Lee, J.Y., Stinson, D.R.: A combinatorial approach to key predistribution for distributed sensor networks. In: IEEE Wireless Communications and Networking Conference, WCNC 2005, New Orleans, LA, USA (2005)
12. Ruj, S., Roy, B.: Key predistribution using partially balanced designs in wireless sensor networks. In: Stojmenovic, I., Thulasiram, R.K., Yang, L.T., Jia, W., Guo, M., de Mello, R.F. (eds.) ISPA 2007. LNCS, vol. 4742, pp. 431–445. Springer, Heidelberg (2007)
13. Ruj, S., Roy, B.: Key predistribution schemes using codes in wireless sensor networks. In: Yung, M., Liu, P., Lin, D. (eds.) Inscrypt 2008. LNCS, vol. 5487, pp. 275–288. Springer, Heidelberg (2009)

14. Al-Shurman, M., Yoo, S.M.: Key pre-distribution using MDS codes in mobile ad hoc networks. In: ITNG, pp. 566–567. IEEE Computer Society Press, Los Alamitos (2006)
15. Steiner, J.G., Neuman, B.C., Schiller, J.I.: Kerberos: An authentication service for open network systems. In: USENIX Winter, pp. 191–202 (1988)
16. Sarkar, P., Saha, A., Chowdhury, M.U.: Secure Connectivity Model in Wireless Sensor Networks Using First Order Reed-Muller Codes. In: MASS 2010, pp. 507–512 (2010)
17. Xu, D., Huang, J., Dwoskin, J., Chiang, M., Lee, R.: Re-examining probabilistic versus deterministic key management. In: Proceedings of the 2007 IEEE International Symposium on Information Theory (ISIT), pp. 2586–2590 (2007)

Comparison of Region Based and Weighted Principal Component Analysis and Locally Salient ICA in Terms of Facial Expression Recognition

Humayra Binte Ali, David M.W. Powers, Richard Leibbrandt, and Trent Lewis

Abstract. With the increasing applications of computing systems, recognizing accurate and application oriented human expressions, is becoming a challenging topic. The face is a highly attractive biometric trait for expression recognition because of its physiological structure and location. In this paper we proposed two different subspace projection methods that are the extensions of basis subspace projection methods and applied them successfully for facial expression recognition. Our first proposal is an improved principal component analysis for facial expression recognition in frontal images by using an extension of eigenspaces and we term this as WR-PCA (region based and weighted principal component analysis). Secondly we proposed locally salient Independent component analysis(LS-ICA) method for facial expression analysis. These two methods are extensively discussed in the rest of the paper. Experiments with Cohn-kanade database show that these techniques achieves an accuracy rate of 93% when using LS-ICA and 91.81% when WR-PCA and 83.05% when using normal PCA. Our main contribution here is that by performing WR-PCA, which is an extension of typical PCA and first investigated by us, we achieve a nearly similar result as LS-ICA which is a very well established technique to identify partial distortion.

1 Introduction

The face has been called the most important perceptual stimulus in the social world [1] with infants as young as three months able to discern facial emotion [2]. This means that the facial expression is a major modality in human face to face communication. From a psychological viewpoint, it is well known that there are seven basic expressions including neutral. In recent years the research in automatic facial expression recognition has been attracted a lot of attention in many fields such

Humayra Binte Ali · David M.W. Powers · Richard Leibbrandt · Trent Lewis
CSEM school, Flinders University
e-mail: ali0041@flinders.edu.au

R. Lee (Ed.): Software Eng., Artificial Intelligence, NPD 2011, SCI 368, pp. 81–89.
springerlink.com   © Springer-Verlag Berlin Heidelberg 2011

as robotics, somnolence detector, surveillance system, physiology etc. The algorithms of Facial expression recognition can be divided into two broad categories:
appearance based approaches and geometrical feature based approaches. Among
appearance based feature extraction, subspace projection techniques are often used
in computer vision problems as an efficient method for both dimension reduction
and finding the direction of the projection with certain properties. Usually, the face
image is considered to lie in a high-dimensional vector space. The subspace projection techniques represent a facial image as a linear combination of low rank basis
images. The popular subspace projection techniques are PCA, ICA and FDA. In the
context of face expression recognition, we attempt to find some basis vectors in that
space serving as much as important directions of projection in a low rank image
subspace. These subspace projection algorithms have been used in Facial Expression Recognition area over the last ten years in the work of [3],[5],[6], [8],[9],[10].
Subspace projection algorithms work by creating low rank basis images and project
the original image as a linear combination of low rank images. By projecting they
employ feature vectors consisting of coefficients of the reduced components. Fig. 1
depicts the algorithm for subspace projection step by step. The paradigm of "recognition by parts" has been popular in the object recognition research area [7]. This is
because local representation is robust to local distortion and partial occlusion. So for
robustness enhancement the basis images should effectively contain a specific important component based local representation that correspond to particular feature
in face space like eyes, lips, nose, eyebrows and facial furrows.

  Although PCA, ICA architecture I, ICA architecture II has been effectively used
in facial expression detection in [3],[5],[6],[8],[9],[10]. Here our main contribution
is to apply part based subspace projection to classify the six basic facial expression
for the reason that the changes in faces due to facial expression is very localized. So
we apply PCA on the upper face and lower face separately and also on the whole

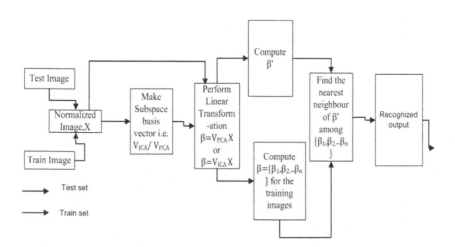

Fig. 1 Subspace Projection Method.

face and by making comparison we propose a weighted PCA (WR-PCA) concept. For some expressions, applying PCA on separate face region outperforms applying PCA on the whole face in terms of facial expression recognition. As for example happiness can be 100% recognized by applying PCA on lower face (lips) only. ICA outperforms PCA but region based PCA gives the nearly same result as ICA architecture I does that we have discussed in the result analysis section. ICA architecture I basis images do not display perfectly local features because the pixels do not belong to locally salient feature regions still have some nonzero weight values although ICA basis images are spatially more localized. These pixel values in non-salient regions would appear as noise and contribute to the degradation of the recognition. So that To overcome this problem we are using here the LS-ICA (locally salient ICA) basis images, where only locally salient feature regions are retained. These regions correspond to important facial feature regions such as eyes, eyebrows, nose, and lips. In [12] researchers successfully used LS-ICA method for face recognition. But here we are using this algorithm for facial expression recognition because of its robustness to partial distortion. There is no work for facial expression recognition based on locally salient ICA.

2 Proposed Algorithm

Here we are proposing two improved subspace projection algorithm and making a comparison between them. The first one is improved PCA which means component based PCA by applying it in partial regions which is our unique contribution. The second one is locally salient ICA method which was proposed by the researchers of [12] and they use it for face recognition. We propose Improved PCA by applying PCA on the partial face and also on the holistic face and do some comparison. Based on the result of different dataset we took the weight factor whether partial or holistic PCA is appropriate for the specific emotion. How we choose the weighting factor please see the improved PCA system section. Next we use the locally salient ICA (LS-ICA) algorithm because of its robustness to partial occlusion and local distortion.

2.1 Region Based Principal Component Analysis

Among many subspace projection techniques, Principal Component Analysis also named as Karhunen Leove Transformation has been widely used in image analysis. Given a set of M training samples we form the image column data matrix $X = (x_1, x_2, ..., x_m)$ and its transpose (image row data matrix) $Y = X^T$. Then we center the data in an image space by first obtaining its mean vector $m = \frac{1}{M} \sum_{j=1}^{m} x_j$. Subtracting the mean vector m from each training vector, that is, $x_j = (x_j - m)$. Now the centered image column data matrix is $x_c = \bar{x}_1, \bar{x}_2, ...\bar{x}_m$. We have to sphere the data using PCA based on the centered observation vector $\bar{x}_1, \bar{x}_2, ...\bar{x}_m.$. The sample covariance matrix is given by

$$Cov_x = \frac{1}{M} \sum_{j=1}^{m} \bar{x}_j \bar{x}_j^T = \frac{1}{M} x_c x_c{}^T \qquad (1)$$

and then the principal components of the covariance matrix are computed by solving $P^T Cov_x P = \prod$ where $\prod$ is the diagonal matrix of eigenvalues and P is the matrix of orthonormal eigenvectors. According to geometry, P is a rotation matrix that rotates the original coordinate system onto the eigenvectors, where the eigenvector associated with the largest eigenvalue is the axis of maximum variance, the eigenvector associated with the second largest eigenvalue is the orthogonal axis with the second largest variance, etc. Then only the β eigenvectors associated with the largest eigenvalues are used to define the subspace, where β is the desired subspace dimensionality.

There are three relevant cause for matching images in the subspace of β eigenvectors. The first is data compression. To compare images in significantly reduced dimensions is more efficient in terms of computation. For example, image vectors with 65,536 pixels (256x256) might be projected into a subspace with only 280x180 dimensions. The second dispute is that the data samples are drawn from a normal distribution. So that axes of large variance probably correspond to signal, while axes of small variance are probably noise. Therefore eliminating these axes improves the accuracy of matching. The third dispute depends on a common pre-processing step, in which the mean value is subtracted from every image and the images are scaled to form unit vectors. Then the projection of images occur into a subspace where Euclidean distance is inversely proportional to correlation between the source images. As outcome, the neighbour matching by measuring Euclidean distance, Angle distance or KNN in eigenspaces those are nearest becomes an efficient approximation to image correlation. Different distance measure algorithm outcomes different outputs.

2.1.1 Proposed System Implementation

We apply PCA on the whole face and on the upper and lower face separately. Then based on the result we add the weight factor with each three separate output and take the value with the greatest weight factor. As we first apply PCA on region(R) based and then we count the weighted(W) matrix we term this method as WR-PCA. The great contribution of this method is that its performance for facial expression analysis is much better than PCA. Another thing to notice that it requires less computation time and less memory than PCA.

$E_{hap} = W_1.U_{hap} + W_2.W_{hap} + W_3.L_{hap}$

$E_{dis} = W_1.U_{dis} + W_2.W_{dis} + W_3.L_{dis}$

$E_{sur} = W_1.U_{sur} + W_2.W_{sur} + W_3.L_{sur}$

$E_{ang} = W_1.U_{ang} + W_2.W_{ang} + W_3.L_{ang}$

$E_{sad} = W_1.U_{sad} + W_2.W_{sad} + W_3.L_{sad}$

$E_{fear} = W_1.U_{fear} + W_2.W_{fear} + W_3.L_{fear}$

Note that for every expression we took the partial output of greatest weight. So the weight of the final recognized expression would be, $W_{expr} = Max(W)$, where $W = W_1 = W_2 = W_3$.

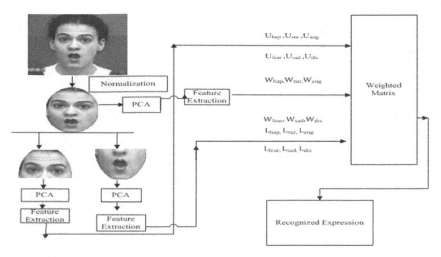

Fig. 2 Proposed Facial Expression Recognition system using WR- PCA.

2.2   *The Locally Salient Independent Component Analysis*

Independent Component Analysis (ICA) is a widely used subspace projection technique which decorrelates the high-order moments of the input in addition to second-order moments. The Locally Salient Independent Component Analysis (LS-ICA) is an extension of ICA in terms of analysing the region based local basis images. In terms of facial expression detection this method outperforms simple ICA due to its ability to identify local distortions more accurately. It needs two step to compute the LS-ICA. First thing is to create component based local basis images based on ICA architecture and then order the obtained basis images in the order of class separability for achieving good recognition.

By imposing additional localization constraint in the process of the kurtosis maximization the LS-ICA method creates component based local basis images. The solution at each iteration step is weighted so that it becomes sparser by only emphasizing larger pixel values. This sparsification contributes to localization. Let V be a solution vector at an iteration step, we can define a weighted solution,W where $W_i = |V_i|^\beta V_i$ and $W = W/||W||$. $\beta > 1$ is a small constant. The kurtosis is maximized in terms of W(weighted solution) instead of V as in equation 2.

$$Kurt(W) = |E(W)^4 - 3(E(b)^2)^2| \qquad (2)$$

By applying equation 3, which is a update rule we can derive a solution for equation 2.

$$S^{(t+1)} = E|V_i|^\beta Q(|V_i|^\beta S^T(t)Q)^3 \qquad (3)$$

Here (in equation 3) S is a separating vector and Q contains whitened image matrix. Then the resulting basis image is $W^i = |V_i|^\beta (S^T Q)_i$. Then according to the algorithm LS-ICA basis images are created from the ICA basis images selected in the decreasing order of the class separability P [11],where $P = \Omega_{between}/\Omega_{within}$. Thus both dimensional reduction and good recognition performance can be achieved. The output basis images contain localized facial features that are discriminant for change of facial expression recognition.

3   Used Datasets and Experimental Setup

For experimental purpose we benchmark our results on three different datasets which are Cohn Kanade Facial Expression dataset[14]. The Cohn Kanade database contain 97 university students and each student has all six basic expressions like happiness, anger, fear, surprise, sad and disgust. For training purpose we use 40 persons images each expressing 6 emotions(40x6=240) and for testing 57 persons images with 6 expressions each.

3.1   Image Normalization and Alignment

First we do the illumination corrections by applying histogram fitting technique. Image sequences are digitized into 280x180 pixel arrays. Only the frontal view image sequences are used here. But some subjects produce little out of-of-plane motion, we do an affine transformation following the work of [13] to normalize face position and to maintain face magnification invariance. Then the normalization of position of all feature points was performed by mapping them to a standard face model based on three facial characteristic points: the medial canthus of both eyes and the uppermost point of the philtrum[13].

Then we make a triangle with three points like medial canthus of both eyes C1 and C2 and the uppermost point of the philtrum P.

Then we find the middlepoint of the triangle ΔC1C2P by applying the algorithm of finding mid-point of a triangle. Then based on the mid-point we half the picture into upper and lower portion.

3.2   Result Analysis

The contribution of the paper is that here we apply three different subspace projection techniques like PCA, WR-PCA and LS-ICA for facial expression recognition on Cohn-Kanade facial expression dataset. Our intention is to analyse which method is suitable for a specific condition. Because its an open question still now as to which subspace projection algorithm is better for different sector of image analysis.

After performing the required normalization we perform the WR-PCA, PCA and LS-ICA. We term our proposed method as WR-PCA because first we are applying PCA on upper and lower region (R) of face and also on whole face and then we took the weighted (W) matrix. Here 40 eigenvectors have been taken for the whole face and 30 eigenvectors for the upper and lower face separately. In terms of distance measure we took a combined distance measure of city-block and mahalanobis distance to get the better performance. We don't intend to evaluate the performance of

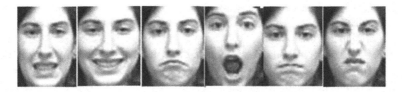

Fig. 3 CKFED(from left: Fear, happy, sad, surprise, disgust, anger)

different distance measure in this paper. When applying PCA on the whole face the average recognition rate is 83.05%.

From Table 1 we can see that some facial expressions are more accurately identified by applying PCA only on lower part or upper part than the whole face. For example here we achieve 100% recognition rate for happiness by performing PCA only on lower face and we achieve 100% recognition rate for surprise by performing PCA on the whole face. For both case we got lower recognition rate by applying PCA only on upper part. This is because for upper part neutral and happiness are confused with each other greatly and also the upper part for surprise and fear are confused greatly. To overcome these short comings we proposed here WR-PCA which we have discussed in the Proposed System Implementation section (Figure 2).

Table 1 Recognition rate of FER based on whole face and region based PCA

PCA	Hap	Ang	Sad	Sur	Fear	Dis
Whole face PCA	91.2%	82.5%	73.7%	100%	84.2%	66.7%
Upper face PCA	52.6%	91.2%	50.9%	94.7%	26.3%	52.6%
Lower face PCA	100%	89.5%	82.5%	96.5%	94.7%	84.2%

By applying our new method we got a recognition rate of 91.81%, a radical increase from the previous one. We got this by applying PCA on upper and lower face and whole face separately and take the weighted sum of these PCA's(WR-PCA). Secondly we applied the LS-ICA to extract the ICs from the dataset and we selected 125 ICs with the selection feature of Kurtosis v alues for each IC for training the model. Now we got an improved recognition rate(93%) by LS-ICA than the result of typical PCA, but nearly same result as region based and weighted PCA (91.81%) that we are proposing here.

Table 2 Recognition rate of FER based on W- PCA and LS-ICA

PCA	Hap	Ang	Sad	Sur	Fear	Dis
Whole face PCA	91.2%	82.5%	73.7%	100%	84.2%	66.7%
WR-PCA	100%	89.5%	82.5%	100%	94.7%	84.2%
LS-ICA	96.5%	93%	82.5%	96.5%	96.5%	93%

88 H.B. Ali et al.

Figure 4 shows the recognition rate of PCA, region based and weighted PCA and LS-ICA. In the graph we use WR-PCA for region based and then weighted PCA that we have proposed here. From the graph its quite clear that the recognition rate of the LS-ICA method was consistently better than that of typical PCA and our proposed region based and weighted PCA regardless of number of dimensions taken. But its performance is nearly similar as WR- PCA. The more interesting thing is that LS-ICA method performs better than the other methods especially at low dimensions. The WR-PCA implementation is computationally less expensive than LS-ICA or typical ICA. So we can derive two conclusions from here: (i) that when computation time is considered WR-PCA is more suitable than LS-ICA because their performance is nearly similar, and (ii) that LS-ICA outperforms the other methods in low dimensions so when a low resolution device is used LS-ICA is more suitable than others.

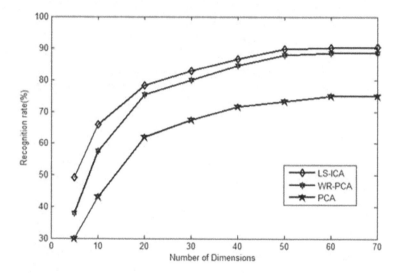

Fig. 4 Performance of recognition of LS-ICA, PCA and WR-PCA in FER

4 Conclusion

The main contribution of the paper is the WR-PCA method that is an extension of typical PCA and first investigated by us. The system is clearly depicted in Figure 2. We successfully implemented WR-PCA on Cohn-Kanade Facial Expression Dataset and achieve a good recognition rate of 91.81% that outperforms typical PCA(83.05% recognition rate) greatly and also this rate is very close to the performance of locally salient ICA(LS-ICA). LS-ICA is a highly suitable method for robustness enhancement in terms of partial distortion and occlusion. We are the first (to the best of our knowledge) to successfully apply LS-ICA in facial expression recognition domain. Our main purpose was to compare among different subspace projection techniques in the area of Facial Expression recognition. Because it's a great challenging topic that which subspace projection technique is suitable for a specific condition.

From the result analysis part we can derive two very important conclusions that when computation time is a concern then WR-PCA is more suitable than LS-ICA because their performance are nearly similar and the computational complexity of ICA is more expensive than PCA. For WR-PCA implementation we need less time and memory than PCA. The second conclusion is that given LS-ICA outperforms the other methods in low dimensionality it is clear then when low resolution device are used then LS-ICA is more suitable than the other methods.

References

1. Frith, U., Baron-Cohen, S.: Perception in autistic children. In: Cohen, D.J., Donnellan, A.M. (eds.) Handbook of Autism and Pervasive Developmental Disorders, pp. 85–102. John Wiley, New York (1987)
2. Charlesworth, W.R., Kreutzer, M.A.: Facial expressions of infants and children. In: Ekman, P. (ed.) Darwin and Facial Expression: A Century of Research in Review, pp. 91–138. Academic Press, New York (1973)
3. Buciu, I., Kotropoulos, C., Pitas, I.: ICA and Gabor representation for facial expression recognition. In: International Conference on Image Processing, vol. 3, pp. II-855–II-858 (2003)
4. Turk, M., Pentland, A.: Eigenfaces for Recognition. Journal of Cognitive Neuroscience (1991)
5. Frank, C., Nöth, E.: Optimizing eigenfaces by face masks for facial expression recognition. In: Petkov, N., Westenberg, M.A. (eds.) CAIP 2003. LNCS, vol. 2756, pp. 646–654. Springer, Heidelberg (2003)
6. Frank, C., Nöth, E.: Automatic pixel selection for optimizing facial expression recognition using eigenfaces. In: Michaelis, B., Krell, G. (eds.) DAGM 2003. LNCS, vol. 2781, pp. 378–385. Springer, Heidelberg (2003)
7. Pentland, A.P.: Recognition by parts. In: IEEE Proceedings of the First International Conference on Computer Vision, pp. 612–620 (1987)
8. Kolenda, T., Hansen, L.K., Larsen, J., Winther, O.: Independent Component Analysis for Understanding Multimedia Content. In: IEEE Workshop on Neural Networks for Signal Processing, pp. 757–766 (2002)
9. Zia Uddin, M., Lee, J.J., Kim, T.S.: An enhanced independent component based human facial expression recognition from video. IEEE Transactions on Consumer Electronics, 2216–2224 (2009)
10. Chen, F., Kotani, K.: Facial Expression Recognition by Supervised independent Component Analysis Using MAP Estimation. IEICE Transactions on Information and Systems, 341–350 (2008)
11. Bartlett, M.S., Movellan, J.R., Sejnowski, T.J.: Face Recognition by Independent Component Analysis. IEEE Trans. Neural Networks 13(6), 1450–1464 (2002)
12. Kim, J., Choi, J., Yi, J.: Face Recognition Based on Locally Salient ICA Information. In: ECCV Workshop BioAW, pp. 1–9 (2004)
13. Lien, J.J.-J., Kanade, T., Cohn, J.F., Li, C.-C.: Automated Facial Expression Recognition Based on FACS Action Units. In: FG 1998, pp. 390–395 (1998)
14. Kanade, T., Cohn, J.F., Tian, Y.: Comprehensive Database of Facial Expression Analysis. In: Proceedings of fourth IEEE International Conference on AFGR, Grenoble, France (2000)

Web Health Portal to Enhance Patient Safety in Australian Healthcare Systems

Belal Chowdhury, Morshed Chowdhury, Clare D'Souza, and Nasreen Sultana

Abstract. The use of a web Health Portal can be employed not only for reducing health costs but also to view patient's latest medical information (e.g. clinical tests, pathology and radiology results, discharge summaries, prescription renewals, referrals, appointments) in real-time and carry out physician messaging to enhance the information exchanged, managed and shared in the Australian healthcare sector. The Health Portal connects all stakeholders (such as patients and their families, health professionals, care providers, and health regulators) to establish coordination, collaboration and a shared care approach between them to improve overall patient care safety. The paper outlines a Health Portal model for designing a real-time health prevention system. An application of the architecture is described in the area of web Health Portal.

1 Introduction

Communication of electronic health information is a vital part of effective healthcare systems. Patient safety and quality of care are two of the nation's most important healthcare challenges [1] today. Current threats to patient safety include a *near miss, an adverse and a sentinel event* [2]. A near miss is a clinical incident that could have resulted in an injury or illness but has been avoided by chance or through timely intervention. It can also draw attention to risks of future events. Near misses are the most common type of clinical incident worldwide including Australia [3]. An *adverse event* is defined as "an unanticipated undesirable or potentially dangerous occurrence in a healthcare organization." [4]. It includes medical errors, medication errors, drug side-effects, patient falls and other injuries, hospital-associated infections, equipment failures and incorrect clinical

Belal Chowdhury · Clare D'Souza
La Trobe University, Melbourne 3086, Australia

Morshed Chowdhury
Deakin University, Melbourne, Victoria 3125, Australia
e-mail: M.Chowdhury@latrobe.edu.au

Nasreen Sultana
Frankston Hospital, Hastings Road, Frankston VIC 3199, Australia

R. Lee (Ed.): Software Eng., Artificial Intelligence, NPD 2011, SCI 368, pp. 91–101.
springerlink.com   © Springer-Verlag Berlin Heidelberg 2011

management [5]. A *sentinel event* is the most serious type of clinical incident. It includes procedures involving the wrong patient or body part, the wrong procedure and surgical instruments left in patients, suicides and death from a medication error. The current challenges of interoperability and patient safety are an important new area of research and development as they pose a greater threat to Australian healthcare. From the quantitative empirical findings in patient safety, researchers found that adverse events and medical errors are common in Australia. According to their analysis, approximately 2-4 per cent of hospital admissions were due to adverse events, with three-quarters of these potentially, preventable. Further, of incidents that occurred in a hospital, 26 per cent were medication related [6]. About 200,000 adverse events (or 5-15 per cent of all patients admitted to hospital) are caused by hospital acquired infections (HAIs), this equates to roughly an additional two million bed days each year in Australia. Up to 70 per cent of HAIs could be prevented if infection control procedures were optimised [7].

The National Health and Hospitals Reform Commission's (NHHRC) recent report found that waste, duplication and inefficiency [8], [9], [10] caused the following impacts:

a) Australians could extend their lives by nearly two years because they missed out on healthcare;

b) unnecessary duplication, high costs, hospital errors and ineffective treatments adversely affected access to healthcare.

c) one in six (about 18 per cent) medical errors (e.g., duplicates pathology and diagnostic tests) are due to inadequate patient information that costs the Australian healthcare system $306 million annually;

d) the incidence of adverse events (i.e. medical errors, medication errors, drug side-effects, patient falls and other injuries, hospital-associated infections, equipment failures and incorrect clinical management) affects about one in six patients;

e) about 2-3 per cent (198,000) of hospital admissions are linked to preventable medication errors that cost the Australian healthcare system $660 million a year;

f) only two-thirds of those who went to emergency departments were seen within clinically appropriate times;

g) hospital errors claim the lives of 4,550 Australians a year; and

h) almost 700,000 hospital admissions cost the Australian healthcare system $1.5 billion a year, and could be avoided, if timely information-sharing and non-hospital healthcare (e.g. community care) had been provided to patients with chronic conditions.

The NHHRC identified that cutting waste and duplication could make Australian hospitals up to 25 per cent more efficient. One-third of patients have to wait longer than the government target of eight hours for a bed in an emergency department. It is estimated that about 1500 deaths each year occur as a result of overcrowding in emergency departments [11]. There are more than 77,000 preventable medication errors per year, of which 27 per cent cause patient harm including at least 7 deaths. It is also estimated that the financial cost of these adverse events for inpatients in Australia is about $2 billion annually. At least 50 per cent of these adverse events are preventable; this would lead to a cost savings of $1 billion (i.e. 3.7 per cent) of the hospitals budget per year.

Another key recommendation by the NHHRC (2009) is the introduction of a Personally Controlled Electronic Health Record (PCEHR) for every Australian, by 2012, to access their health-related information. Currently, Australian States and Territories hold varying types of patient data on different databases that cannot be viewed by other healthcare providers or relevant government agencies. Almost 98 per cent of GP's (General Practitioner) computers and software contains patient information for their own records in Australia, but connection and compatibility between healthcare providers (hospitals, physicians, GPs, pathology labs and allied health groups) does not yet exist [12]. There is very little current national data collection on patient care and hospital performance. While physicians in Australian States and Territories keep their own records, they do not have access to their patient's information outside their offices. Also healthcare providers rarely have access to these patient records and emergency room doctors know little about patient's pre-existing conditions.

The analysis of quantitative empirical findings in patient safety also demonstrated that the Australia's growing and ageing population faces numerous challenges due to the poor access to patient medical information, lack of coordination and collaboration between care providers and huge increases in demand for better healthcare. Mistakes (such as wrong medication) in medical care can occur anywhere in the healthcare system – at hospitals, doctor's surgeries, nursing homes, pharmacies, or patients' homes. This also indicates the urgent requirement for long-term care for the aging population, efficient facilities and more resources. These changes are possible by the use of Information and Communication Technologies (ICT) and emerging technology (such as web Health Portal) as they have all the potential to empower the healthcare industry to provide better care.

The Health Portal is capable of enabling a patient to obtain a wide variety of health services remotely through the Internet [13] as it makes an entire range of new technologies available to care providers (e.g. hospitals) to facilitate communications and build relationships in healthcare systems. Health professionals, care providers, and patients can take advantage of using the Health Portal to improve the flow of information during an episode of care, where doctors can access a patients' medical records more easily, get immediate access to lab test results, deliver prescriptions directly to pharmacists and empower patients [14], [16], [18]. When consumers are better informed, they make better decisions, enhance health performance and improve overall welfare. The Australian Federal Government has recently announced the implementation of PCEHR by 2012 to: reduce avoidable hospital admissions, undertake chronic disease management, prevent medication errors or mismanagement (adverse healthcare incidents) and avoid the duplication of tests and procedures (as test results are not available between healthcare settings or there is a lack of care coordination among them).

The paper is structured as follows: section 2 outlines the Health Portal model. Section 2.1 provides a broad discussion of healthcare identifiers service. Section 3 outlines the architecture of the Health Portal. Section 4 illustrates the web-based

application of Health Portal. Section 5 illustrates the practical implications of the
Health Portal. Section 6 concludes the paper.

2 Health Portal Model

The Health Portal is capable of bringing electronic communication between patients and health services together [15]. The penetration of portals in health has
been gaining momentum in Australia, because of their ultimate integration into
Electronic Health Records (EHR). The Federal Government recently enacted legislation to develop Healthcare Identifiers Service (HI Service) and its recent
budget (2010-2011) moves to spend $466.7 million over two years to launch na-
tional systems of PCEHR [16].

The web Health Portal in this research is mainly used to view patient clinical
test (for example, pathology and radiology) results, discharge summaries, obtain
prescription renewals, referrals, schedule appointments and carry out physician
messaging to enhance the information exchanged, managed and shared in the Australian healthcare sector. It will move away from paper-based systems to the use of
electronic access, transmission and recording of health information. Additionally,
it will enable remote patient learning for preventative care and disease management. Every Australian will be able to choose to have an EHR and access his/her
health information through an intranet and extranet like managing bank accounts
online [16]. The main functionality of a Health Portal is shown in Fig. 1.

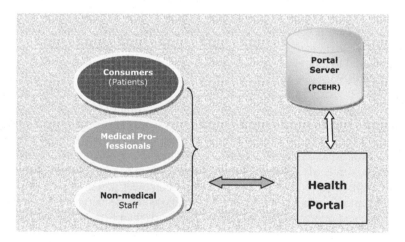

Fig. 1 Main components of the Health Portal

The Health Portal involves all the authorized users (consumers, individual care
providers and employees of health organizations) of the medical records accessing
remotely in real-time from a centralized portal server or database (i.e., PCEHR)
that stores the patient medical information such as patient history, medications, lab
test results, prescriptions, and referrals. The authorized users have to sign up with

the Health Portal service provider (i.e. Medicare Australia). The Portal helps health professionals to take a direct and significant approach to treat a particular patient quickly as the doctor is able to see the patients' entire medical record. It has different access levels for users such as patients, and health providers (both individual and organization). Each time a record from PCEHR is accessed, the details of the user and time records an audit log [17]. The users obtain a user name and password to obtain access to the relevant sections in the Health Portal. The data to be stored and accessed through the portal can be stored on-site or remotely hosted with the consent of the legitimate user.

2.1 Healthcare Identifiers (HI) Service

The Australian Federal, State and Territory governments have developed a national HI Service to improve the security and efficient management of personal health information within a framework of strict privacy laws. The HI has been designed for use by individual (patient), healthcare providers and employees of healthcare organisations. The HI service uniquely identifies individuals and health providers and manages the allocation, maintenance and use of HI. This service will improve individuals and health providers' with confidence that the right health information is associated with the right individual at the point of care. There are three types of Healthcare Identifiers:

a) *Individual Healthcare Identifier (IHI)*: a unique mandatory health ID number (16 digit electronic health numbers) that will be allocated by the next financial year (2011-2012) to all consumers (Australian citizens, permanent residences) enrolled in Medicare and others (visitors) who seek healthcare in Australia [17].

b) *Healthcare Provider Identifier – Individual (HPI-I)*: allocated to health providers (medical specialists, GPs, health professionals employed with public health services, allied health professionals such as Optometrists, speech therapists, social workers) involved in providing patient care.

c) *Healthcare Provider Identifier – Organisation (HPI-O)*: allocated to organisations (state health departments, hospitals, medical practices, pathology or radiology laboratories, phamacies) that deliver healthcare.

The HI service consists of the following:

1) a national database which contains the unique 16 digit healthcare identification numbers. This database only stores HI information (patient's name, address and date-of-birth). It will not store any clinical data and it is separate to a PCEHR.

2) a healthcare provider directory which lists contacts, locations and service details for healthcare providers and organisations, and

3) access controls which provides secure online access to authorised users.

As the operator of the HI service, Medicare Australia is responsible for ensuring the security of the national database and assigning healthcare identifiers to healthcare providers and individuals [17]. Medicare Australia will automatically assign an IHI to individuals who are enrolled for Medicare benefits or who have a Department of Veterans' Affairs (DVA) treatment card. A Medicare number is not an IHI, as it is not unique. Some people have more than one Medicare number as they are members of more than one family and may be on multiple cards.

If a person is not registered with Medicare, but seeks healthcare, he/she will be assigned a temporary identifier by the HI service as authorized by legislation rather than individual consent. Authorised healthcare providers will be able to obtain a unique identifier from the HI Service for their patient records without obtaining the patient's consent. HIs are an important building block to enable a national PCEHR system to be introduced in 2012, which will use HIs to facilitate the identification of a patient and healthcare provider. With the consent of patients, medical information will be stored electronically in a highly secure environment for later instantaneous retrieval by the authorised users. The PCEHR will provide [16] the following:

a)   Patients' health information including medical history, medications, lab test results, referrals, discharge summaries, prescriptions, allergies and immunisations;
b)   Secure and effective access of information remotely for both patients and healthcare providers via the internet ; and
c)   Rigorous governance and oversight to maintain privacy.

3 Health Web Portal Architecture

The architecture of a Health Portal is depicted in Fig. 2. The left part of the figure shows the architecture of connectivity services (for example, Web Health Portal

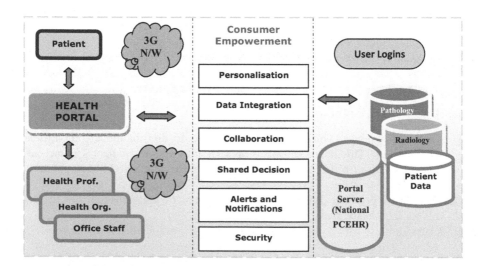

Fig. 2 The architecture of a Health Portal

for patients, health professionals and health organisations) to integrate any health related data such as patient data, pathology data and radiology data. The middle part shows all the patient-centric functionalities and services of a portal. Finally, the right hand side of the architecture corresponds to the user login and national portal server or database, which includes all patients' clinical data and other information. In the healthcare setting, care providers need to carry out their daily responsibility of providing quality care to their patients. As health professionals are highly mobile throughout their day, they often require access to a number of systems and also need to access their office remotely for scheduling follow-up appointments and procedures to drive efficient clinical workflow and optimize patient care via a wired or wireless network (for example, 3G using pocket PC, PDA, iPhone and smart phone).

4   Health Portal Application

Fig. 3 shows the snapshot of a home page of a Health Portal application where health professionals, health organisations and patients can connect and access their medical information. The application is developed using PhP (a server-side scripting language), MySQL, HTML and JavaScript technologies. This portal is a stand-alone web site and is connected to the proposed integrated PCEHR. This Health Portal enables patients to communicate and cooperate with their doctors and care providers in an easy and secure way. All these pages are designed to provide better interaction between the patients, health professionals, care providers and patient medical information via the Internet.

The main function of this application is to assist health professionals in accessing patient personalized latest information (dashboard) to diagnose at the point of care, or remotely, for effective and efficient services as shown in Fig. 3(a) and 3(b).

Fig. 3 Home page of a Health Portal application

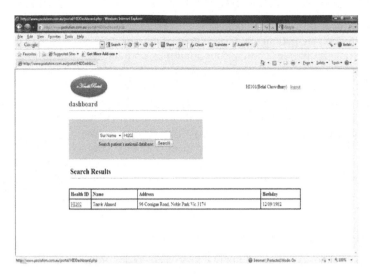

Fig. 3(a) Health professional's dashboard

Fig. 3(b) Health professional's dashboard

This dashboard also assists health professionals keep a track of the health condition of patients, receiving alerts of deteriorating patients and contacting patients via SMS, and email services.

The Patient Dashboard screen allows patients to view the latest medical information and other features such as managing their own account and refill requests. It allows patients to schedule appointments with physicians and can also to request prescription renewals. This page also assists patients to update their personal details as shown in Fig. 3(c).

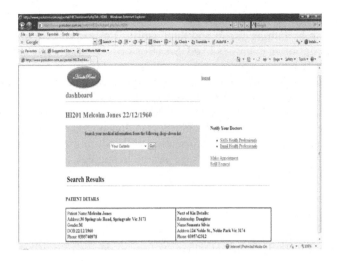

Fig. 3(c) Patient's dashboard

Similarly, the health organisation dashboard screen provides health organisations with an opportunity to search and view their patient's details from the national database at the point of care or remotely in real-time to maximise the effective care.

5 Practical Implications of the Health Portal

The implementation of a Health Portal will promote preventive care. It will provide patients with an opportunity to control what medical information is fed into their EHR and who can access it online, as around 75 per cent of Australian population now use the Internet. It will allow patients' to view their updated medical information in real-time from anywhere in Australia. This will eventually boost patient safety, improve healthcare delivery, bring about better clinical decision support and overcome waste and duplication. Additionally, accurate and accessible information allows patients to become better educated regarding their diseases, become actively involved in the decision making of their own health, which could radically change the experience some patients have with their doctors, eventually leading to activities such as electronic health. In addition, patients will no longer have to remember the details of their medical and care history and will not have to retell them every time they meet their care providers. Patients will also be able to make an appointment, request prescription refills, send and receive messages to and from the doctor or nurse via SMS and email and complete health assessments.

The Health Portal will facilitate health providers to access on demand services such as appointments, modify or update any patient's information on their national database, literally anytime from anywhere in Australia. The electronic exchange of information by health providers can deliver immediate benefits for patients using IHIs (user identification) in areas like prescriptions, referrals, discharge summaries and lab tests to improve patient safety, improve healthcare delivery, and cost saving by cutting waste and duplication. For example, the implementation of e-prescription in countries like USA (Boston), Denmark and Sweden where they are currently enjoying reduced costs, and saving time to improve productivity per prescription by over 50 per cent. Due to the effective access of patient information for both physicians and test ordering, and a results management system through e-referral, USA and France have reduced over 70 per cent and Denmark over 50 per cent of the time spent by physicians chasing up test results. Also providing timely and accurate patient information will reduce medical and medication incidents, avoid unnecessary tests, save scarce resources and improve patients overall care in the Australia health system [16].

6 Conclusions and Future Work

In this paper we have described architecture for the design of a web Heath Portal. We have shown the application and implementation of the above system. Using a Health Portal, health professionals and care providers have a chance to view patient's personalized latest health information to diagnose at the point of care to improve patient safety, improve healthcare delivery and make cost savings by cutting waste and duplication. The application also helps patients to view the latest medical information and other features such as managing their own account, making doctor appointments and refill requests.

Further study is needed in Health Portals to evaluate more advanced portal services such as video conferencing, online chat, access to medical reports, online bill payments and their impact on patient satisfaction, and quality of care.

The Health Portal can help patients' better understand and manage the effects of their treatment by providing educational materials such as health information. Facilitating discussions of confidentiality and health will be an important direction for future research.

References

1. IOM (Institute of Medicine), To Err is Human: Building a Safer Health System. National Academy Press, Washington, DC (November 1999)
2. JCI (Joint Commission International), Understanding and Preventing Sentinel and Adverse Events in Your Health Care Organization, Illinois, USA (2008)
3. Pearson, D.D.R.: Patient Safety in Public Hospitals, Victorian Auditor-General report, PP No 100, Session 2006-08 (2008) ISBN 1 921060 68 9
4. Grundmann, H.: Emergence and resurgence of methicillin-resistant Staphylococcus aureus as a public health threat. Lancer 368(874), 230 (2006)
5. NSW Health, NSW Health Patient Survey 2007 Statewide Report, Sydney (2008)
6. Runciman, W.B.: The Safety and Quality Of Health Care: Where Are We Now? Shared meanings: preferred terms and definitions for safety and quality concepts. MJA (The Medical Journal of Australia) 184(suppl. 10), S41–S43 (2006)

7. Cruickshank, M., Ferguson, J.: Reducing harm to patients from health care associated infection: the role of surveillance. Australian Commission on Safety and Quality in Health Care (July 2008)
8. Bita, N.: Sensitive health records to be stored on national database, The Australian (May 12, 2010),
   http://www.theaustralian.com.au/in-depth/budget/
   sensitive-health-records-to-be-stored-on-national-
   database/story-e6frgd66-1225865267860 (accessed on June 20, 2010)
9. NHHRC (The National Health and Hospitals Reform Commission), Beyond the blame game: accountability and performance benchmarks for the next Australian healthcare agreements, Canberra, p. 5 (2009)
10. Metherell, M.: Hospital botches kill 4500, The Age (July 27, 2009),
    http://www.theage.com.au/national/
    hospital-botches-kill-4500-20090726-dxjm.html
    (accessed on January 27, 2010)
11. Forero, R., Hillman, K.M., McCarthy, S., Fatovich, D.M., Joseph, A.P., Richardson, D.B.: Access block and ED overcrowding. Emergency Medicine Australasia 22(2), 119–135 (2010)
12. Gearin, M.: National health system to go online, Australian Broadcasting Corporation (October 13, 2009)
13. Meijer, W.J., Ragetlie, P.: Empowering the patient with ICT-tools: the unfulfilled promise. Stud. Health Technol. Inform. 127, 199–218 (2007)
14. Fensel, D.: Ontologies. A Silver Bullet for Knowledge Management and E-Commerce, 2nd edn. Springer, Heidelberg (2003)
15. WHO (World Health Organisation), eHealth: proposed tools and services (2005),
    http://apps.who.int/gb/ebwha/pdf_files/EB117/
    B117_15-en.pdf (accessed on August 01, 2010)
16. Ross, Which new technologies do you think we should be using to improve aged care? Department of Health and Ageing (2010),
    http://www.yourhealth.gov.au/internet/yourhealth/
    publishing.nsf/content/
    blogtopic?Opendocument&blogid=BLOG00015 (accessed on July 19, 2010)
17. Medicare Australia, Healthcare Identifiers Service Information Guide, p. 1–3 (2010),
    http://www.medicareaustralia.gov.au (accessed on July 12, 2010)
18. The Australian, E-health can bring $28bn in benefits: business council (November 23, 2009)

Peer-Based Complex Profile Management

Mark Wallis, Frans Henskens, and Michael Hannaford

Abstract. The rising popularity of Web 2.0 applications has seen an increase in the volume of user-generated content. Web Applications allow users to define policies that specify how they wish their content to be accessed. In large Web 2.0 applications these policies can become quite complex, with users having to make decisions such as 'who can access my image library?', or 'should my mobile number be made available to 3rd party agencies?'. As the policy size grows, the ability for everyday users to comprehend and manage their policy diminishes. This paper presents a model of policy configuration that harnesses the power of the Internet community by presenting average-sets of policy configuration. These policy "profiles" allow users to select a default set of policy values that line up with the average case, as presented by the application population. Policies can be promoted at an application level or at a group level. An XML approach is presented for representing the policy profiles. The approach allows for easy profile comparison and merging. A storage mechanism is also presented that describes how these policies should be made persistent in a distributed data storage system.

1 Introduction

Security and privacy are key concerns as the Internet continues to evolve. The introduction of Web 2.0 [5] has seen an increase in web applications that rely upon user-generated content. When a user uploads content to a web application s/he is relying on that application to protect the data, and to only use it in ways to which the user has agreed. Social Networking websites often provide a "profile" [7] that can be configured to describe how a user wishes the web application to use their data. As the complexity of these sites increases, the profiles are becoming larger and hence more difficult to configure. Users currently have two options available to them in regards to configuring their profiles:

Mark Wallis · Frans Henskens · Michael Hannaford
Distributed Computing Research Group, University of Newcastle, Australia
e-mail: mark.wallis@uon.edu.au

R. Lee (Ed.): Software Eng., Artificial Intelligence, NPD 2011, SCI 368, pp. 103–111.
springerlink.com © Springer-Verlag Berlin Heidelberg 2011

- Rely on the default values provided by the web application.
- Review every configuration value manually and choose an appropriate setting.

The first option requires the user to rely on the web application provider to select a default value that is in the best interest of the end user. This has been proven to not always be a reliable option [4]. The second option is a labour-intensive task that relies on the end user understanding each presented profile option and selecting an appropriate answer. As the size of the profile configuration grows, there is an increased tendency for users to "give up" and not complete the full configuration. This leaves many options at their default settings and the user in the same situation as those who take the first option.

There is a clear need for solution that allows quick, easy configuration of profile options, while also allowing users the ability to tweak options to their specific needs. Every user of a web application implementing such configurable user profiles has the same policy creation problem; it therefore makes sense to leverage the user population to provide a solution. This paper presents an approach that relies on surveying user profiles to create an average set of profile responses. This average set is summarised at multiple levels of the user hierarchy, from connected groups to an application-wide view. These profile averages are presented as pre-determined "policies", and allow the user to rely on average case configuration data based on what their peers have previously configured.

This paper provides advances in the following areas:

- An XML language definition for representing and storing policy sets.
- A process for averaging and comparing policy sets in a hierarchical manner.
- A storage platform for user policies based on a distributed storage mechanism.
- A mechanism for detecting policy structure change, including the creation of new policy options.
- Methods of enforcing policy configuration options, based upon a centralised storage mechanism, at a company level.

The remainder of this paper begins by presenting a framework for storing policy information. Section 2 reviews previous research in this area. Sections 3 and 4 deal with the way policy is created and distributed between the web application and the end user. Section 5 presents a distributed storage model for policy sets and associated meta-data. Sections 6 and 7 define the way security is enhanced by this solution, and how group management fits into the design. Finally, Sections 8, 9 and 10 present a proof-of-concept implementation and summarise the advances presented in this paper.

2 Previous Work

The benefits that Social Networking can contribute to various security and privacy functions have previously been investigated. For example Besmer, et.al [1] provide a recent survey of the ways input from social networks can affect policy configuration choices made by end users. Their research suggests that, in particular, the visual prompting must be sufficiently strong for any recognisable effect to be effected. The complexity of policy configuration in Social Networking applications has also been previously presented [3], along with reviews on the importance of users' awareness of the affects of privacy policy configuration [6].

This previous research led the authors to recognise the importance of privacy configuration, with an emphasis on the need for a solution that is scalable, easy to use, and secure. Previous work in this area does not provide a complete technical solution, nor does it address issues such as policy storage, or how such policies can assist with overall data security. This paper presents a technical solution to these problems.

3 Profile XML

The first step in automating policy configuration is to define a method for storing "profiles". A profile is a set of policy answers specific to an entity, an entity being a user, or a group of users. The various questions that need answering by a particular policy are specific to each web application. These questions take the form of meta-data related to the profile.

The Web Application would first make an XML document, describing the options that need to be specified by a user's policy, publicly available. An example of such meta-data is given below:

```xml
<profile_meta>
 <application>
  <name>Social Networking Website</name>
   <address>www.social.com</address>
 </application>
 <policy>
  <privacy>
   <option name="MobilePhoneAccess">
    <answer name="Public" desc="Public Access"/>
    <answer name="Group" desc="Friends Access"/>
    <answer name="Private" desc="Private Access"/>
   </option>
   <option name="EmailAccess">
    <answer name="Public" desc="Public Access"/>
    <answer name="Group" desc="Friends Access"/>
    <answer name="Private" desc="Private Access"/>
   </option>
  </privacy>
 </policy>
</profile_meta>
```

Entity-specific profiles matching this meta-data are then stored using matching XML, as presented below:

```
<profile>
  <entity>
    <entity_name>Mark Wallis</entity_name>
    <entity_type>User</entity_type>
  </entity>
  <application>
    <name>Social Networking Website</name>
    <address>www.social.com</address>
  </application>
  <policy>
    <privacy>
      <option name="MobilePhoneAccess" value="Public"/>
      <option name="EmailAccess" value="Group"/>
    </privacy>
  </policy>
</profile>
```

The above XML format is scalable to suit multiple policy options grouped in a hierarchical manner. Ensuring that all entities store their profile options in the same way provides benefits when it comes to comparing and averaging profile data. For example, a specific profile can be validated against profile meta-data to ensure that all question/answer elements are provided. This allows applications to upgrade and 'version' their profile meta-data, while retaining backwards compatibility with existing user profiles. By way of demonstration, a web application may choose to add a new privacy configuration option as shown below:

```
<profile_meta>
  <policy>
    <privacy>
      <option name="HomeAddressAccess">
        <answer name="Public" desc="Public Access"/>
        <answer name="Private" desc="Private Access"/>
      </option>
    </privacy>
  </policy>
</profile>
```

A simple SAX parse of the application's presented profile meta-data, could be used to compare it to the user profile. This would show that the profile was missing a (Profile/Policy/Privacy/Option@name=HomeAddressAccess) element, and hence was out-of-date compared to the current policy information being requested by the application.

While XML has been chosen as the storage mechanism in this specific implementation, the concept is generic and could equally be implemented in other languages, such as JSON.

4 Profile Creation and Operations

Now that a way of storing profile information has been defined we, must provide a way for the user to create their personal profile XML.

To achieve manual creation of profiles, the web application would present a user interface that collects profile information from the user. This information is then made persistant in the XML format described above. Such a manual solution is based on the typical solution we currently see used by Social Networking applications. As previously discussed, manual profile creation is unwieldy, so the solution proposed by this paper allows the user to auto-generate a profile based on profiles averaged from the population base of the application.

Previous work presented by Besmer, et al [1] introduces visual cues that indicate the most common answer to each policy question to the end user. Our XML schema approach allows these common suggestions to be auto-generated at multiple levels. For instance, the user may wish to select defaults based on the entire population, or perhaps based only on the average generated when surveying the "friends" in their social network. The definition of a typical, or 'majority average' view is highly user-group specific.

Using an XML language allows for advanced operations such as merging of, and comparison between, profiles. XPath [2] operations can be used to extract singular values from multiple policy sets, average the results, and generate a new policy using the average value. Entities can compare polices, using basic XML comparison techniques, to highlight differences in their configuration. Strict enforcement of XML schema will allow multiple concurrent sequential parsers to efficiently compare numerous policies without a large memory footprint. By way of comparison, current systems that rely solely on user-interfaces for policy definition, do not give end users access to the above style of operations.

A key benefit of this XML approach focuses on the meta-data. Web Applications will, as time goes on, need to change the policy information they collect from their users. For example, a new feature may be added that introduces a new policy option. Such change enacts a change to the profile meta-data XML document, and because of the document structure it is easy to compare the meta-data to a user profile, and to see which new options need to be completed by the user. Again, the user may choose to simply accept a default for any newly added options.

5 Profile Storage

Profile meta-data is generated and stored by web applications themselves. According to the current Web 2.0 design, user profile information is also stored by web applications. This is sub-optimal for a number of reasons, including:

- Users have no protection against web application owners changing policy settings without their (the users') knowledge.
- Comparison, and other profile operations, must be implemented by the web application.

- End users are not able to look at their profile configuration in any way other than that supported and allowed by the web application.

These limitations demonstrate the benefits of offloading the storage of profile information to an entity controlled by the end-user. User-controlled distributed storage models [9] provide a platform for storing the profile information at a data storage provider chosen by the end user.

In a distributed content storage system (DSS) the user manages, or procures, a data storage service that is used to store all their user-generated content. Web Applications 'subscribe' to pieces of information held in these data stores, and offload the responsibility for storing and securing the data to the data owner. Enhanced client software dynamically builds pages out of framework code provided by the web application, combined with data provided by numerous distributed content storage systems. This model solves issues relating to data ownership, data freshness and data duplication. The user can rely on their own data storage service being the single-version-of-the-truth for all their data [9].

Accordingly, web applications such as social networking websites would request read-only access to a user's profile information via a web service interaction with that user's data storage service. Any updates to policy meta-data would be pushed from the web application to the storage service using a publish/subscribe [10] operation, at which time the end-user would be notified that their policy needed review. Figure 1 shows how the distributed storage solution works in relation to the end user, and to the web application.

When profile information is stored in DSS, the fact that the storage is controlled by the end-user removes the risk that web application owners may change policy information without the user's consent. Objects requested from the distributed data storage service may be cached, with updates available using a publish/subscribe mechanism. This allows the web application to request a user's profile upon user login, without having to constantly re-request the full profile each time a policy setting needs to be evaluated.

6 Data Security

The same DSS used to store the user's profile is used to store the user's generated content, such as their image library and contact details. Additional policy security can be implemented at the DSS level. For instance, if a user stores their mobile phone number in the DSS, and their policy permits only private access to this data, the DSS can enforce the policy at either the data linkage or data access stages, so that a badly-implemented (or rogue) web application is not able to act against the user's wishes.

7 Group Management

For profiles to be aggregated at multiple levels, we need a way to store a user hierarchy. The storage location of this hierarchy will greatly influence which component of the overall system is responsible for generating and storing aggregated profile data.

The use of a distributed data service arrangement for storing profiles ensures that no one data store (which might otherwise have, as a side effect, performed aggregation) is responsible for provision of all user profile data. Web applications will continue to access in-memory cached versions of profiles, which they could use to maintain running averages across their respective user populations. For example, each time a user logged in, data could be added to the average set stored locally by the web application. The result is that more frequent users of the web application will have greater influence on the average data set, while the impact of users who have not logged into the application for long periods of time would eventually become negligible.

Alternatively, a web application that required aggregation based on the entire (rather than recently active) user base, could regularly request policy information directly from the users' respective DSSs for the purpose of aggregation. It should be noted, however, that this is not scalable for large user populations.

Another possibility is that the DSS could store the user hierarchy data. Continuing with the social networking example, it is clear that the way a particular user

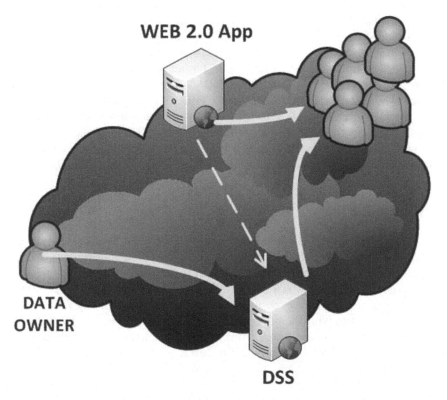

WEB 2.0 App

DATA OWNER

DSS

Fig. 1 Distributed Profile Storage Model

relates to others in the user community represents a valuable piece of information. If the user hierarchy were stored in the distributed store, it could become the DSS's responsibility to query the DSSs of the user's friends to generate the average sets. This distributed architecture leads itself to a more component-based approach [8] to web application development.

8   Implementation

A proof-of-concept implementation of the XML profile schema, and associated operations, has been implemented within previously-reported systems relating to distributed content storage [9]. XML schemas were persisted on a user-managed storage service, and accessed via a web service from a prototype address book application. The following sample policy options were defined:

- Mobile Phone Access - private, group or public
- Email Address Access - private, group or public

Operations were created to summarise those values at both group and an application wide levels. Users were then given the option to automatically accept defaults at either level, and then to alter those settings if they wished.

This implementation demonstrated the feasibility of removing responsibility for policy storage from the web application. Caching was implemented within the web application and distributed storage framework, thus ensuring that the policy file was only fetched from the DSS on user login. This reduced any performance impact that would otherwise be introduced through DSS fetches every time a policy-based decision was made by the web application. Policy data storage was also able to take advantage of existing DSS features, such as push notifications. This allowed for policy updates to occur mid-session, if required. Use and observation of the implementation indicated no adverse performance implications associated with the benefits afforded by the combination of distributed, XML-expressed profiles.

9   Conclusion

This paper presents an XML-based technique for representing user policy information. The structure of this representation allows for efficient operations, such as comparisons between policies, and averaging of policy sets.

By separating user policy storage from Web 2.0 applications, we added a level of user protection against service providers' changing and evolving policy without appropriately notifying their users. The storage of policy information in a user-specified distributed storage server allows policy information to be applied across multiple web applications. This provides the users with the possibility of a 'single version of the truth' for all their policy decisions.

The presented techniques add overall data security to a user's data, while also addressing the issues of scalability and performance. The way user relationships (or 'connections') are stored in Web 2.0 systems, is also discussed.

10 Observations and Future Work

Storage of multiple policies in a user controlled storage service creates the possibility of a unique identifier that identifies a user across multiple services. For example, at present a user would have a unique ID with his/her social networking website, and that would differ from the unique ID used by his/her photo library website. Centralised cross-application profile management suggests that the storage server should generate and propagate a single identifier for each user across the multiple systems accessed by the user.

Privacy concerns with such a model obviously need to be reviewed. In the absence of such a centrally provided unique identifier, we are seeing the ID used by popular social networking websites (e.g. Hotmail or Facebook login) becoming de-facto choices for unique identifiers - with smaller websites allowing you to 'log in' using your social networking profile. The effect of this, including the automatic propogation of security profiles, warrants further investigation.

References

[1] Besmer, A., Watson, J., Lipford, H.R.: The impact of social navigation on privacy policy configuration. In: Proceeding of the Sixth Symposium on Usable Privacy and Security (2010)

[2] Clark, J., DeRose, S.: Xml path language (xpath) (1999), W3C
http://www.w3c.org/TR/1999/REC-xpath-19991116/

[3] Lipford, H.R., Besmer, A., Watson, J.: Understanding privacy settings in facebook with an audience view. In: Proceedings of the 1st Conference on Usability, Psychology and Security (2008)

[4] McKeon, M.: The evolution of privacy on facebook (2010), Personal Website,
http://www.mattmckeon.com/facebook-privacy/

[5] O'Reilly, T.: What is web 2.0. O'Reilly Net (2005),
http://www.oreillynet.com/pub/a/oreilly/tim/news/
2005/09/30/what-is-web-20.html

[6] Strater, K., Lipford, H.R.: Strategies and struggles with privacy in an online social networking community. In: Proceedings of the 22nd British HCI Group Annual Conference on People and Computers: Culture, Creativity, Interaction - Volume 1 (2008)

[7] Vickery, G., Wunsch-Vincent, S.: Participative Web And User-Created Content: Web 2.0 Wikis and Social Networking. Organization for Economic (2007)

[8] Wallis, M., Henskens, F.A., Hannaford, M.R.: Component based runtime environment for internet applications. In: IADIS International Conference on Internet Technologies and Society, ITS 2010 (2010a)

[9] Wallis, M., Henskens, F.A., Hannaford, M.R.: A distributed content storage model for web applications. In: The Second International Conference on Evolving Internet, IN-TERNET 2010 (2010)

[10] Wallis, M., Henskens, F.A., Hannaford, M.R.: Publish/subscribe model for personal data on the internet. In: 6th International Conference on Web Information Systems and Technologies, WEBIST 2010, INSTICC (2010)

Detecting Unwanted Email Using VAT

Md. Rafiqul Islam and Morshed U. Chowdhury

Abstract. Spam or unwanted email is one of the potential issues of Internet security and classifying user emails correctly from penetration of spam is an important research issue for anti-spam researchers. In this paper we present an effective and efficient spam classification technique using clustering approach to categorize the features. In our clustering technique we use VAT (Visual Assessment and clustering Tendency) approach into our training model to categorize the extracted features and then pass the information into classification engine. We have used WEKA (www.cs.waikato.ac.nz/ml/weka/) interface to classify the data using different classification algorithms, including tree-based classifiers, nearest neighbor algorithms, statistical algorithms and AdaBoosts. Our empirical performance shows that we can achieve detection rate over 97%.

Keywords: Email, classification, TP, FP, Spam.

1 Introduction

Spam is defined as unsolicited commercial email or unsolicited balk email, and it has become one of the biggest worldwide problems facing the Internet today. The Internet is becoming an integral part of our everyday life and the email has treated a powerful tool intended to be an idea and information exchange, as well as for users' commercial and social lives. Due to the increasing volume of unwanted email called as spam, the users as well as Internet Service Providers (ISPs) are facing enormous difficulties. The cost to corporations in bandwidth, delayed email, and employee productivity has become a tremendous problem for anyone who provides email services.

Email classification is able to control the problem in a variety of ways. Detection and protection of spam emails from the e-mail delivery system allows end-users to regain a useful means of communication. Many researches on content based email classification have been centered on the more sophisticated classifier-related issues [10]. Currently, machine learning for email classification is an

Md. Rafiqul Islam · Morshed U. Chowdhury
School of Information Technology,
Deakin University, Melbourne, Australia
e-mail: {rislam,muc}@deakin.edu.au

R. Lee (Ed.): Software Eng., Artificial Intelligence, NPD 2011, SCI 368, pp. 113–126.
springerlink.com   © Springer-Verlag Berlin Heidelberg 2011

important research issue. The success of machine learning techniques in text categorization has led researchers to explore learning algorithms in spam filtering [1, 2, 3, 4, 10, 11, 13, and 14]. However, it is amazing that despite the increasing development of anti-spam services and technologies, the number of spam messages continuously increasing.

Due to the rapid growth of email and spam over the time, the anti-spam engineers need to handle with large volume email database. When dealing with very large-scale datasets, it is often a practical necessity to seek to reduce the size of the dataset, acknowledging that in many cases the patterns that are in the data would still exist if a representative subset of instances were selected. Further, if the right instances are selected, the reduced dataset can often be less noisy than the original dataset, producing superior generalization performance of classifiers trained on the reduced dataset. In our previous work we have reduced the data set by ISM (Instance Selection Method) to select a representative subset of instances, enabling the size of the new dataset to be significantly reduced [32]. We used k-mean clustering technique to select the cluster leader (centeriod) and reduce the size based on distance measure [32]. In this paper we propose email classification using VAT clustering technique to categorize the email features. The main focus of this paper is to reduce the instances of the email corpora from training model which are less significant in relation to the classification. Our empirical evidence shows that the proposed technique gives better accuracy with reduction of improper instances from email corpora. The rest of the paper is as follows: section 2 will describe the related work; section 3 will describe the cluster classification methods, section 4 will describe the proposed email classification architecture and its detail description. Section 5 presents the key findings. Finally, the paper ends with conclusion and references in section 6 and 7 respectively.

2   Related Works

2.1 Classification Technique

In recent years, many researchers have turned their attention to classification of spam using many different approaches. According to the literature, classification method is considered one of the standard and commonly accepted methods to stop spam [10].

Classification based algorithms are commonly use learning algorithms. Given that classification algorithms outperform other methods, when used in text classification (TC) [10] and other classification areas like biometric recognition and image classification [10,12,17], researchers are also drawn to its uses for spam filtering.

The email classification can be regarded as a special case of binary text categorization. The key concepts of classification using learning algorithms can be categorized into two classes and there are N labelled training examples: $\{x_1, y_1),....,(x_n, y_n), x \in \mathfrak{R}^d$ where d is the dimensionality of the vector [13].

Many classification methods have been developed with the aid of learning algorithms such as Bayesian, Decision Tree, K-nn (K-nearest neighbour), Support Vector Machine (SVM) and boosting. All these classifiers are basically learning methods and adopt sets of rules. Bayesian classifiers are derived from Bayesian Decision Theory [2,8]. This is the simplest and most widely used classification method due to its manipulating capabilities of tokens and associated probabilities according to the user's classification decisions and empirical performance.

We have selected five algorithms, SVM, NB (Naïve Bayes), DT (Decision Tree), RF (Random Forest) and WEKA IB1 in our experiment due to its simplicity and observing their empirical as well as analytical performance as well as representing the spectrum of major classification techniques available. We also apply AdaBoost to each of these to determine if indeed this improved our results. However, it is very difficult to select a good classifier which can always focus on desired output space with ease of training model [12]. The data we obtained, shows that not all algorithms performed equally well; however, in each case, AdaBoost was able to improve classification performance.

2.2 Cluster Classification

Traditional Clustering Methods

Clustering techniques can be naturally used to automatically identify the unwanted software or emails [20,21,22]. However, clustering is an inherently difficult problem due to the lack of supervision information. Different clustering algorithms and even multiple trials of the same algorithm may produce different results due to random initializations and random learning methods. The well-known clustering algorithm   are K-means [24] and K-medoids [23,25] which assigns a set of samples into clusters using an iterative relocation technique [23].

In [27], they proposed a cluster ensembles technique using combination of different base clustering techniques for making ensemble method. In [26], authors proposed a parameter-free hybrid clustering algorithm (PFHC) which combines the hierarchical clustering and K-means algorithms for malware clustering. PFHC evaluates clustering validity of each iteration procedure and generates the best K value by comparing the values.

Clustering Using VAT

The VAT algorithm displays an image of reordered and scaled dissimilarity data. The VAT approach is based on pair wise dissimilarity relationships between n objects in a given set of object or relational data. VAT presents the pair wise dissimilarity information as a square digital image with n^2 pixels, after the objects are suitably reordered; the image is better able to highlight potential cluster structure.

VAT was initially used by [28]. The author used this approach for visually assessing cluster tendency using ordered dissimilarity images. Their algorithm was linked with Prim's algorithm to find the minimal spanning tree of a weighted graph. The approach was able to signal the presence of well separated clusters via the manifestation of dark blocks of pixels on the main diagonal of the ordered dissimilarity image (ODI). Their technique is also applicable to all dimensions and all numerical data types, complete or incomplete.

The authors in [18], proposed improved VAT called iVAT which is a method to combine a path-based distance transform with VAT. This technique can overcome the limitations of VAT which is inability to highlight cluster structure of a pair wise, dissimilarity matrix, when the matrix structure contains highly complex clusters. They have used a wide range of primary and comparative experiments on synthetic and real-world data sets to show the effectiveness of their algorithms.

Further work on VAT algorithm by [28], developed an approach for visually assessing of cluster tendency for the objects represented by a relational data matrix.

In [29], authors proposed a modified technique which can significantly reduces the computational complexity of the iVAT algorithm [18]. Their formulation reduces the complexity from $O(N^3)$ to $O(N^2)$ for producing both the VAT and iVAT images.

3 Proposed Model for Email Classification

This section presents the proposed email classification model based on cluster classification technique. We have used VAT process for clustering and WEKA interface to classify the VAT generated outputs using well known classifiers. The general approach of our model is to extract a broad set of features from each email sample that can be passed to our classification engine. Figure 1 shows the architecture of classification system.

3.1 Description of the Model

We aim to detect the spam using machine learning techniques and propose a method capable of automatically classifying spam. An outline of our learning approach is given by the following basic steps:

- Email pre-processing
- Feature extraction (FE)
- Clustering using VAT
- Classification and evaluation

3.1.1 Pre-processing Incoming Email

The objective of the email transformation is to pre-process the email messages into a homogeneous format that can be recognized by the classifier. In this process, we collect the feature of the incoming email contents and convert it into a vector space where each dimension of the space corresponds to a feature of whole corpus of the email message. The initial transformation is often a null step that has the output text as just the input text. Character-set-folding, case-folding and MIME (Multipurpose Internet Mail Extensions) normalization are used for initial transformation in many cases. A corpus of emails is being used in our system are taken from [10].

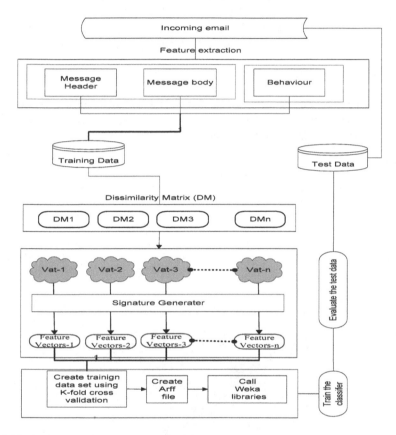

Fig. 1 Block diagram of our proposed model.

3.1.2 Feature Extraction

Our general approach to the classification problem is to extract a broad set of features from each email sample that can be passed to our classification engine. In our feature extraction process we have used three set of information from each email, they are: 1) message body, ii) message header and iii) email behaviour.

Each email is parsed as text file to identify each header element to distinguish them from the body of the message. Every substring within the subject header and the message body was delimited by white space. Each substring is considered to be a *token*, and an *alphabetic word* was also defined as a token delimited by white space. The alphabetic word in our experiments was chosen only English alphabetic characters (A-Z, a-z) or apostrophes. The tokens were evaluated to create a set of 21 hand-crafted features from each e-mail message. The study investigates the suitability of these 21 features in classifying unwanted emails.

The behavioural features are included for improving the classification performance, in particular for reducing false positive (FP) problems. The behavioural features in our experiment consists the frequency of sending/receiving emails, email attachment, type of attachment, and size of attachment and length of the email.

3.1.3 Clustering the Data

After feature extraction, we use the data set to make dissimilarity matrix to create VAT image. The following Figure 2 shows a sample dissimilarity matrix (R) and corresponding dissimilarity image in Figure 3.

$$
R = \begin{pmatrix}
0 & 0.73 & 0.19 & 0.71 & 0.16 \\
0.73 & 0 & 0.59 & 0.12 & 0.78 \\
0.19 & 0.59 & 0 & 0.55 & 0.19 \\
0.71 & 0.12 & 0.55 & 0 & 0.74 \\
0.16 & 0.78 & 0.19 & 0.74 & 0
\end{pmatrix}
$$

Fig. 2 Dissimilarity matrix

Fig. 3 Dissimilarity image

After generating the VAT cluster image, we order the dissimilarity matrix (R) and then we again create the VAT image. The following Figure 4 shows the ordered dissimilarity matrix (R') generated from Figure 2 and their corresponding VAT image in Figure 5.

$$R' = \begin{pmatrix} 0 & 0.12 & 0.59 & 0.73 & 0.78 \\ 0.12 & 0 & 0.55 & 0.71 & 0.74 \\ 0.59 & 0.55 & 0 & 0.19 & 0.19 \\ 0.73 & 0.71 & 0.19 & 0 & 0.16 \\ 0.78 & 0.74 & 0.19 & 0.16 & 0 \end{pmatrix}$$

Fig. 4 Ordered dissimilarity matrix R'

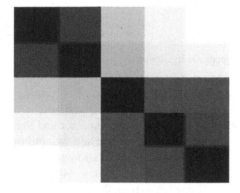

Fig. 5 VAT image from R'

3.1.4 Feature Selection

From the output of VAT cluster, we select the cluster leader (centeroid of the cluster) of each cluster and map the data using distance matrix. We then select the features based on closest distance from the centeroid. For selecting the closest distance we use the formula presented in our previous work [31]. We also reduce the redundant and noisy features from our data set based on data cleaning technique using WEKA tools. The following block diagram shown in Figure 6 is illustrating the feature selection technique.

3.1.5 Learning and Classification

Machine learning techniques are applied in our model to distinguish between spam and legitimate emails. The basic concept of the training model of a classifier is that it employs a set of training examples of the form $\{(x_1, y_1), ..., (x_k, y_k)\}$ for the projection of a function $f(x)$. Here the values of x are typically of the form $< x_{i,1}, x_{i,2}, ..., x_{i,n} >$, and are composed of either real or discrete values. The y values represent the expected outputs for the given x values, and are usually drawn from a

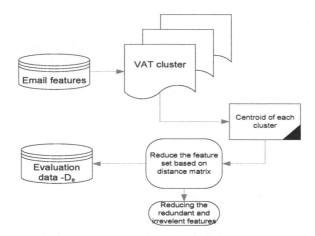

Fig. 6 Block diagram of feature selection process

discrete set of classes. Consequently, the task of a learning model involves the approximation of the function $f(x)$ to produce a classifier.

In the training model, we use a particular data set and the same number of instances by randomly selected from other sets. Then we split the data in K different fold for making training set. In our classification process we use the value of K=10 for making k-fold cross validation, that is 90% for training and 10% for testing. Then we use the training set to train the classifier and evaluate the classification result by using the test data. The same rotating process applies for all other sets.

In our classification process, we built an interface program with WEKA for data classification. The interface program allows the system to select the data sets and the corresponding classifiers according to the testing requirements rather than the default in WEKA tools.

3.1.6 Evaluation Matrix

The performance of the classifier algorithms using supervised learning algorithms is typically evaluated using 'confusion matrix'. A confusion matrix contains information about actual and predicted classifications produced by a classification system. The following Table 1 shows the confusion matrix for a two class classifier.

Table 1 Confusion matrix of two class classifier

Predicted Actual	Positive (P)	Negative (N)
Positive (P)	TP	FP
Negative (N)	FN	TN

The four cells of Table 1 contain the counts for true positive (TP), false positive (FP), true negative (TN) and false negative (FN) respectively. Some common standard metrics have been defined from Table 1, such as precision, recall and accuracy.

The matrices used in our testing evaluation process are outlined below:

- False Positive (FP). A FP refers to when a classifier incorrectly identifies an instance being positive. In this case a false positive is when a legitimate email is classified as a spam message.
- False Negative (FN). A FN is when a classifier incorrectly identifies an instance as being negative. In this case a false negative is when a spam message is classified as being a legitimate email.
- Precision (Pr). The precision is the proportion of the predicted positive cases that were correct, as calculated using the equation $P = TP / (TP + FP)$
- Recall (Re). This measures the portion of the correct categories that were assigned, as calculated using the equation $R = TP / (TP + FN)$
- Accuracy (Acc). This measures the portion of all decisions that were correct decisions, as calculated using the equation $A = (TP + TN)/(TP + FP + TN + FN)$

4 Empirical Evidence

4.1 Experimental Setup

This section describes the experimental setup and the methodology of our proposed classification system. In our experiment, we have tested five classifiers, NB, SVM, IB1, RF and DT.

The methodology used in the experiment is as follows:

Step 1 Extract the features from incoming emails

Step 2 Generate dissimilarity matrix
  a. Combine the features from all categories and make an integrated data file.
  b. Split the file into two groups; i) Training (90%) and test (10%)
  c. Create dissimilarity matrix R of training file base on distance measure
  d. Generate the corresponding dissimilarity image.

Step 3 Generating the VAT cluster
  a. Ordered the dissimilarity matrix (R')
  b. Create the VAT image from the R'
  c. Create centeroid – leader of the cluster, and labeling the instances of each cluster.

Step 4 Generate the features vector; create global list of features after selecting the features from each VAT cluster.

Step 5 Classification and evaluation
    a.   Select the classifier Ci $[i=1..n]$
    b.   Use K fold cross validation (K=10)
    c.   Call WEKA libraries to train the Ci using Training Data
    d.   Evaluate the test data
    e.   Repeat until k=0
    f.   Repeat for other classifiers
    g.   Repeat for other data sets

4.2 Experimental Results

This section presents the classification outcome of different algorithms. Table 2 presents the average of experimental results according to the classifiers. It has been shown that all the classifier accuracy is almost similar except the naïve bayes algorithm which is worst compared to others. The RF shows best performance among the classifiers.

Table 2. Average classification results

Algorithm	FP	FN	Pr	Re	Acc
NB	0.044	0.011	0.942	0.922	0.936
SVM	0.03	0.12	0.96	0.95	0.962
IB1	0.08	0.04	0.94	0.954	0.952
DT	0.0	0.09	0.968	0.95	0.962
RF	0.0	0.24	0.97	0.963	0.977

Figure 7 shows the classifiers accuracy compared to existing techniques [32]. The present evaluation shows almost similar performance with trivial variation. The classifier SVM and IB1 are bit lower compared to existing method, however the RF and NB show best performance. In comparison to the above, the present work shows better performance, that is ~ 98 (97.7) % accuracy (RF classifier), compared to our previous work [11,12,32].

4.3 ROC Report for Multi-tier Classification

Table 3 shows average performance of the experiment using Receiver Operating Characteristic (ROC) report. We have used four important measurement values in this ROC report, AUC (Area Under Curve) estimation values, AUC estimated standard error (StdErr), a 95% of CI (Confidence Interval) for both lower and upper limits, and 1-sided Probability-values (P-value).

    The AUC is a popular measure of the accuracy of an experiment. The larger the AUC, the better the experiment is at predicting by the existence of the classification. The possible values of AUC range from 0.5 (no diagnostic ability) to 1.0

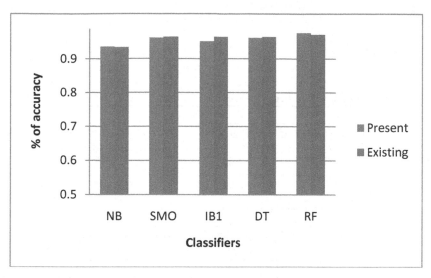

Fig. 7 Comparison of classification accuracy with existing techniques [32]

(perfect diagnostic ability). The CI option specifies the value of alpha to be used in all CIs. The quantity (1-Alpha) is the confidence coefficient (or confidence level) of all CIs. The P-value represents the hypotheses tests for each of the criterion variables.

A useful experiment should have a cut-off value at which the true positive rate is high and the false positive rate is low. In fact, a near-perfect classification would have an ROC curve that is almost vertical from (0,0) to (0,1) and then horizontal to (1,1). The diagonal line serves as a reference line since it is the ROC curve of experiment that is useless in determining the classification.

Table 3 The ROC report of five classifiers

ROC Estimation	NB	SVM	IB1	DT	RF
AUC	0.93514	0.96988	0.96248	0.96948	0.97222
AUC (StdErr)	0.02015	0.01791	0.02591	0.01891	0.01936
95% of CI	0.77645	0.84627	0.81527	0.83827	0.81421
	0.97625	0.99440	0.98340	0.9899	0.99613
P-value	(<0.001)	(<0.001)	(<0.001)	(<0.001)	(<0.001)

Table 4 presents the cost-benefit analysis (CBA) for the ROC report. The CB ratio is the ratio of the net cost when the condition is absent, to the net cost when it is present. In this experiment we select the cut-off value for which the computed

Table 4 Cost benefit analysis with prevalence =0.00001 for four classifiers

	Cutoff Value	Sensitivity	Specificity	Cost - Benefit When Ratio = 0.5000	Cost - Benefit When Ratio = 0.7000	Cost - Benefit When Ratio = 0.8000	Cost - Benefit When Ratio = 0.9000
NB							
	-1.000	1.0000	0.2846	-35769.4048	-50077.5667	-57231.6476	-64385.7286
	1.000	0.4122	0.9562	-2191.2857	-3067.9649	-3506.3045	-3944.6441
SVM							
	-1.000	1.0000	0.3296	-33517.5897	-46925.0256	-53628.7436	-0332.4615
	1.000	0.4122	0.9863	-686.2969	-960.9806	-1098.3224	-1235.6642
IB1							
	-1.000	1.0000	0.3296	-33517.5897	-46925.0256	-53628.7435	-60332.4615
	1.000	0.4499	0.9863	-686.2593	-960.9429	-1098.2848	-1235.6266
DT							
	-1.000	1.0000	0.3296	-33517.5897	-46925.0256	-53628.7436	-60332.4615
	1.000	0.4129	0.9863	-686.2962	-960.9799	-1098.3217	-1235.6635
RF							
	-1.000	1.0000	0.3296	-33517.5897	-46925.0256	-53628.7436	-60332.4615
	1.000	0.4129	0.9863	-686.2962	-960.9799	-1098.3217	-1235.6635

cost value is maximized (or minimized) and the prevalence is considered (=0.00001), which is the actual probability of the condition in the population.

5 Conclusion and Future Work

This paper presents and effective classification technique to identify the unwanted emails from user mailbox, based on clustering method using VAT into the training model. We have reviewed different classification algorithms and have considered five classifiers based on our simulation performance. Our empirical performance shows that, we achieved overall classification accuracy ~98%, which is significant compared to existing techniques [11, 12, 32]. In future work we have a plan to extract the features from dynamic information from real time emails contents and pass to our classification engine to achieve better performance.

References

[1] Zhang, J., et al.: A Modified logistic regression: An approximation to SVM and its applications in large-scale text categorization. In: Proceedings of the 20th International Conference on Machine Learning, pp. 888–895. AAAI Press, Menlo Park (2003)

[2]   Sahami, M., Dumais, S., Heckerman, D., Horvitz, E.: A bayesian approach to filter-
      ing junk e-mail. In: Learning for Text Categorization: Papers from the Workshop,
      Madison, Wisconsin, AAAI Technical Report WS-98-05 (1998)
[3]   Androutsopoulos, I., et al.: Learning to filter spam e-mail: A comparison of a Naive
      Bayesian and a memory-based approach. In: Proceedings of the Workshop on Ma-
      chine Learning and Textual Information Access, 4th European Conference on Prin-
      ciples and Practice of Knowledge Discovery in Databases, Lyon, France, pp. 1–13
      (2000)
[4]   Drucker, H., Shahrary, B., Gibbon, D.C.: Support vector machines: relevance feed-
      back and information retrieval. Inform. Process. Manag. 38(3), 305–323 (2003)
[5]   Islam, R., Chowdhury, M., Zhou, W.: An Innovative Spam Filtering Model Based on
      Support Vector Machine. In: Proceedings of the IEEE International Conference on
      Intelligent Agents, Web Technologies and Internet Commerce, vol. 2, pp. 348–353
      (28-30, 2005)
[6]   Cohen, W., Singer, Y.: Context-sensitive learning methods for text categorization.
      ACM Transactions on Information Systems 17(2), 141–173 (1999)
[7]   Cristianini, N., Shawe-Taylor, J.: An introduction to Support Vector Machines and
      other kernel-based learning methods. Cambridge University Press, Cambridge (2000)
[8]   Kaitarai, H.: Filtering Junk e-mail: A performance comparison between genetic pro-
      gramming and naïve bayes, Tech. Report, Department of Electrical and Computer
      Engineering, University of Waterloo (November 1999)
[9]   Huan, L., Lei, Y.: Toward Integrating Feature Selection Algorithms for Classification
      and Clustering. IEEE Transaction on Knowledge and Data Engg 17(4) (2005)
[10] Islam, R., Zhou, W.: An innovative analyser for multi-classifier email classification
      based on grey list analysis. The Journal of Network and Computer Applica-
      tions 32(2), 357–366 (2008)
[11] Islam, M.R., Chowdhury, M.: Spam filtering using ML Algorithms. In: Proceedings
      of the WWW/Internet Conference, Portugal (2005)
[12] Islam, M.R., Zhou, W.: Architecture of adaptive spam filtering based on machine
      learning algorithms. In: Jin, H., Rana, O.F., Pan, Y., Prasanna, V.K. (eds.) ICA3PP
      2007. LNCS, vol. 4494, pp. 458–469. Springer, Heidelberg (2007)
[13] Islam, M., Chowdhury, M., Zhou, W.: Dynamic feature selection for spam filtering
      using support vector machine Support Vector Machine. In: 6th IEEE International
      Conference on Computer and Information Science (ICIS 2007), Melbourne, Austral-
      ia, July 11-13 (2007)
[14] Chong, P.H.J., et al.: Design and Implementation of User Interface for Mobile Devic-
      es. IEEE Transactions on Consumer Electronics 50(4) (2004)
[15] Hunt, R., Carpinter, J.: Current and New Developments in Spam Filtering. In: IEEE
      International Conference on Networks, ICON 2006, vol. 2, pp. 1–6 (2006)
[16] Wang, X.-L., Cloete, I.: Learning to classify email: a survey. In: IEEE ICMLC 2005,
      vol. 9, pp. 5716–5719. IEEE, Los Alamitos (2005) Isbn:0-7803-9091-1
[17] Eleyan, A., Demirel, H.: Face Recognition using Multiresolution PCA. In: IEEE In-
      ternational Symposium on Signal Processing and Information Technology, pp. 52–55
      (2007)
[18] Wang, L., Nguyen, U.T.V., Bezdek, J.C., Leckie, C.A., Ramamohanarao, K.: iVAT
      and aVAT: Enhanced visual analysis for cluster tendency assessment. In: Zaki, M.J.,
      Yu, J.X., Ravindran, B., Pudi, V. (eds.) PAKDD 2010. LNCS, vol. 6118, pp. 16–27.
      Springer, Heidelberg (2010)

[19] Wang, Y., Ye, Y., Chen, H., Jiang, Q.: An improved clustering validity index for determining the number of malware clusters. In: 3rd International Conference on Anti-Counterfeiting, Security, and Identification in Communication, ASID 2009 (2009)

[20] Bayer, U., Comparetti, P.M., Hlauschek, C., Kruegel, C., Kirda, E.: Scalable,behavior-based malware clustering. In: NDSS 2009 Security Symposium (2009)

[21] Lee, T., Mody, J.J.: Behavioral classification. In: EICAR 2006 (May 2006)

[22] Bailey, M., Oberheide, J., Andersen, J., Mao, Z.M., Jahanian, F., Nazario, J.: Automated classification and analysis of internet malware. In: Kruegel, C., Lippmann, R., Clark, A. (eds.) RAID 2007. LNCS, vol. 4637, pp. 178–197. Springer, Heidelberg (2007)

[23] Xu, R., Wunsch II, D.: Survey of Clustering Algorithms. IEEE Transactions on Neural Networks 16, 645–678 (2005)

[24] Hartigan, J., Wong, M.: A k-means clustering algorithm. Applied Statistics 28, 100–108 (1979)

[25] Kaufman, L., Rousseeuw, P.J.: Finding groups in Data: An Introduction to Cluster Analysis. Wiley, New York (1990)

[26] Xue, Z., Feng, H.S., Ye, Y., Jiang, Q.: A Parameter-Free Hybrid Clustering algorithm used for Malware Categorization. In: ASID 2009, pp. 480–483 (2009)

[27] Ye, Y., Tao, L.: Automatic Malware Categorization Using Cluster Ensemble. In: SIGKDD (2010)

[28] Bezdek, J.C., Hathaway, R.J.: VAT: A Tool for Visual Assessment of (Cluster) Tendency. In: Proc. Int'l Joint Conf. Neural Networks, pp. 2225–2230 (2002)

[29] Haven, T.C., Bezdek, J.C.: An Efficient Formulation of the Improved Visual Assessment of Cluster Tendency (iVAT) Algorithm. In: IEEE Transactions on Knowledge and Data Engineering, IEEE Computer Society Digital Library. IEEE Computer Society, Los Alamitos (2011)

[30] Bezdek, J.C., Hathaway, R.J., Huband, J.M.: Visual Assessment of Clustering Tendency for Rectangular Dissimilarity Matrices. IEEE Transactions on Fuzzy Systems 15(5) (October 2007)

[31] Kate, M.S., Islam, R.: Meta-learning for data summarization based on instance selection method. In: WCCI 2010, Spain, July 18-23 (2010)

[32] Islam, R., Xiang, Y.: Email Classification Using Data Reduction Method, Chinacom (August 25-27, 2010)

Improving Smart Card Security Using Elliptic Curve Cryptography over Prime Field (F_p)

Tursun Abdurahmonov, Eng-Thiam Yeoh, and Helmi Mohamed Hussain

Abstract. This paper describes the use of Elliptic Curve Cryptography (ECC) over Prime Field (F_p) for encryption and digital signature of smart cards. The concepts of ECC over prime field (F_p) are described, followed by the experimental design of the smart card simulation. Finally, the results are compared against RSA algorithms.

1 Introduction

The use of smart cards has grown tremendously in the past decade. Practically all daily activities can be converted to use smart cards as a means to submit personal data into different types of information systems. Many examples can be found such as financial, passport, travel, and health care systems [2, 6].

A major concern for these smart card systems is the security of the data stored in the smart cards. Access to data in smart cards by unauthorized parties may result in financial losses. Thus it is vital that the data in smart cards are secured through reliable cryptography algorithms such as Elliptic Curve Cryptography (ECC) and Secure Hash Algorithm (SHA). These methods are used to encrypt and decrypt the data in smart cards and generate digital signatures for verification of smart cards.

2 Smart Card Cryptography

Cryptography uses mathematic methods of sending messages in disguised form so that only the intended recipients can remove the disguise and read the message. In Figure 1.1 cryptography scheme of smart card system is portrayed as hierarchy of cryptography. Cryptography algorithms of smart card system [2, 6] are used for encryption and decryption, digital signature and key exchange. They can be categorized as public key (Asymmetric key), private key (Symmetric key), and hash functions (Secure Hash Algorithm (SHA)) [1, 17].

Tursun Abdurahmonov, Eng-Thiam Yeoh, and Helmi Mohamed Hussain
Multimedia University, Faculty of Information Technology,
Jalan Multimedia, 63100, Cyberjaya, Selangor, Malaysia
e-mail: {tursun.abdurahmono07,etyeoh,helmi.hussain}@mmu.edu.my

R. Lee (Ed.): Software Eng., Artificial Intelligence, NPD 2011, SCI 368, pp. 127–140.
springerlink.com   © Springer-Verlag Berlin Heidelberg 2011

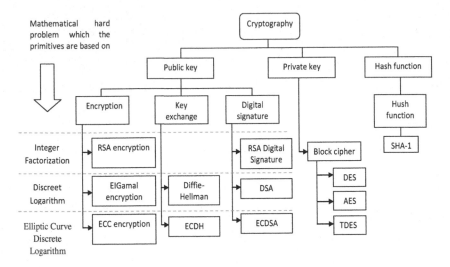

Fig. 1 Scheme of smart card cryptography

There are three objectives of cryptography on the smart card which are [16, 6]:

- Confidentiality: secrecy of messages, which are kept the data into integrate circuited chip that involves in en electronic transaction, actually is provided by encryption.
- Data integrity: all data of the smart card have not been altered and destroyed in an unauthorized manner, which is typically provided by digital signature.
- Authentication: the recipient can verify that the received message has not been changed in the course of being transmitted, which is supplied by digital signature.

2.1 Encryption

Encryption is a process in which the sender encrypts the message in such a way that only the receiver will be able to decrypt the message [7]. Encryption in smart card systems can use the cryptographic algorithms shown in Figure 1.

With Private Key cryptosystem (symmetric key), sender and receiver use one private key or common key. Private Key cryptosystem are used in smart card systems are defined in standards such as DES (Data Encryption Standard), TDES (Triple DES) and AES (Advanced Encryption Standard [2, 7]. A major advantage is that private key cryptosystems are faster than public key cryptosystems. Private key cryptosystem processes are illustrated in Figure 2.

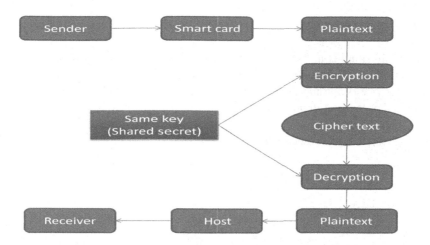

Fig. 2 Private key crypto encryption scheme

Public key cryptography (asymmetric key) was introduced in 1975 by Diffie, Hellman, and Merkle to address the aforementioned shortcomings of symmetric key cryptography [4, 9, 16]. In contrast to private key schemes, public key schemes require two keys, which are a public key and a private key. Advantage of public key cryptosystems is the higher security compared to private key cryptosystems. However, public key cryptosystems are usually significantly slower than private key cryptosystems. Figure 3 illustrates the public key crypto encryption scheme.

The implementation of public key cryptosystems in the smart card systems uses Rivest Shamir Adleman (RSA) public key cryptography such as RSA encryption, RSA digital signature and RSA key excange [9, 16]. Nowadays, Elliptic Curve Cryptography (ECC) public key cryptosystems are emerging as alternative RSA public key cryptosystems in the smart card technologies because ECC public key cryptosystem [8, 9, 10, 18] has the same security level with RSA, small key length, lower power consumption, fast computation, and small key bandwidth.

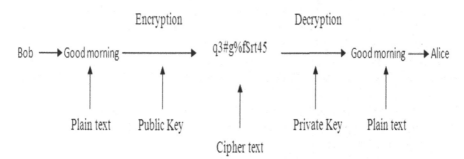

Fig. 3 Public key crypto encryption scheme

2.2 Digital Signature

Digital signature is an electronic analogue of a hand-written signature and estab-
lishes both of sender authenticity and data integrity assurance. Thus, digital signatures are one of the most interesting and original applications of modern cryptography that the first digital signature idea was proposed by Diffie and Helman in
1976 [4]. Figure 4 shows the digital signature process using public key cryptosystems (RSA) and hash functions (SHA).

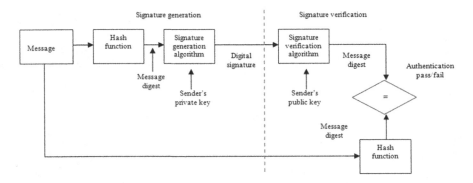

Fig. 4 Digital signature scheme

The digital signature process uses private and public key for signature generation and verification. The hash functions are used for authentication process.

2.3 Secure Hash Algorithm

A hash function is a mathematical algorithm that produces a unique reference value based on a data value. A hash algorithm for a digital signature takes variable
length of data (may be thousands or millions bits) and produces a fixed length
output. Hash function algorithms are part of symmetric algorithms that are a widely deployed in cryptography for message authentication.

Nowadays, SHA hash functions are very famous algorithms in smart card systems because SHA functions are very secure. Popular versions of SHA hash functions are SHA-1 and SHA-2 (SHA-224, SHA-256, SHA-384, and SHA-512)
[1, 17].

The SHA hash functions generate the output by transforming the input data in
80 steps, which applies a sequence of logical functions in each step. Figure 5
illustrates one of the steps.

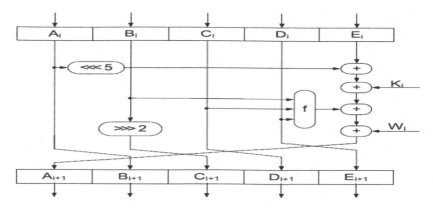

Fig. 5 One step of the state updates transformation of SHA-1

3 Elliptic Curve Cryptography

ECC was proposed in 1985 by Neil Koblitz [5] and Victor Miller [14]. ECC can be used to provide a digital signature, key exchange, and encryption. ECC is based on the difficulty of an underlying mathematical problem, such as Discrete Logarithm Problem (DLP) but is the strongest public-key cryptographic system known [12, 13, 16] to date as well as it is emerging as an attractive public key cryptosystem.

3.1 ECDLP

The foundation of every cryptosystem is a hard mathematical problem that is computationally infeasible to solve. The discrete logarithm problem (DLP) is the basis for the security of many cryptosystems including the Elliptic Curve Cryptosystem. More specifically, the ECC relies upon the difficulty of the Elliptic Curve Discrete Logarithm Problem (ECDLP). The ECDLP is districted over the points of an elliptic curve [11, 15, 19]. Since the ECDLP appears to be significantly harder than the DLP (Discrete Logarithm Problem), the strength-per-key bit is substantiality greater in elliptic curve systems than in conventional discrete logarithm system [16].

The core of elliptic curve arithmetic is an operation, which is called scalar multiplication. Scalar multiplication computes with the following equation:

$$Q = k * P \tag{1}$$

Point P in the Elliptic Curve E is given; minimum positive integer n is following the equation $nP = O$, where O is point at infinity (∞). Integer n is called the order of the point P and n is a divisor of the order of the curve E. Equation (1) is a subgroup of points $\langle P \rangle$ where $\langle P \rangle$ is a finite cyclic group:

$\langle P \rangle = \{P, 2P, 3P, 4P, ..., (n-1)P, nP\}$. The problem of calculating k from equation (1.1) is called discrete logarithm problem over elliptic curve.

Elliptic Curve $E(F_q)$, point $P \in E(F_q)$ of order n and point $Q \in E(F_q)$, integer $k \in [0, (n-1)]$ are determined in equation (1.1). From equation (1), Q is a public key, and k is a private key. Thus, ECDLP is based on the intractability of scalar multiplication of points on elliptic curves. Note that in equation (1) the number k is called "the discrete *log* of Q base P," and point P is called the base point in this problem.

In Equation (1), it is computationally difficult to calculate k that is a scalar multiplier in the equation. The security of ECC relies on the hardness of solving the ECDLP based on this. From a finite cyclic group, order n is a large prime that makes ECDLP becomes harder. Elliptic curve has some points (base point G), which have a large prime order n, and $\#E(F_q) = n * P$.

3.2 Finite Field

Finite field (F_{p^n}) consists of a finite set of elements F together with two binary operations on F such as addition and multiplication binary operation that satisfy the arithmetic properties. Finite field is called Galois Field (in honor of Evariste Galois) in some sources [16]. The order of the finite field is the number of elements in the field. Finite field is based on q and is denoted F_q. So that, elliptic curves are based finite field such as Galois Fields $(GF(p))$ and $(GF(2^m))$: $GF(p) = F_p$ and $GF(2^m) = F_{2^m}$. There are two finite fields, F_p that the first finite field $q = p$ an odd prime finite field where F_p is prime field. The second finite field is $q = 2^m$ for some $m \geq 1$ which is called F_{2^m} binary field. Consequently, this finite field involves elliptic curve finite field: they are elliptic curve over prime field (F_p) and elliptic curve over binary field (F_{2^m}).

3.3 The Group Law

The group law is a very important part to account arithmetic explanation of ECC, which is based on point addition and point doubling that is the difference in ECC over prime field (F_p) [15, 16, 18]. The different equations are to account point addition and point doubling so that, the group law is completely unlike. The group law of ECC over prime field (F_p) is explained in Table 1.

Figure 6 and Figure 7 are drawn geometric explanation of point addition and point doubling.

3.4 Scalar Multiplication

Scalar multiplication is the heart of ECC over prime field(F_p). Scalar multiplication is called point multiplication because coordinate systems are based on points. Scalar multiplications is also significant operation, have chosen to implement ECC over prime field (F_p) in smart card systems, was used the binary method while it does not require pre-computation, and uses less memory than other efficient methods [16, 9]. Main equation of scalar multiplication is given as follows:

$$Q = [k]P = P + P + \cdots + P \qquad (2)$$

Table 1 Group law of ECC over prime field (F_p)

ECC over prime field F_p
$y^2 = x^3 + ax + b$
Identity: $P + \infty = \infty + P$
$P \in E(F_p)$
Negatives
$P = (x, y) \in E(F_p)$
$(x, y) + (x, -y) = \infty$
$P_1 = (x_1, y_1), P_2 = (x_2, y_2), Q = (x_3, y_3) \in E(F_p)$
Point addition $(P_1 \neq P_2)$
$P_1 + P_2 = Q = (x_3, y_3)$
$\gamma = (y_2 - y_1)/(x_2 - x_1)$
$x_3 = \gamma^2 - x_1 - x_2$
$y_3 = \gamma(x_1 - x_3) - y_1$
Point doubling $(P_1 = P_2)$ but $(P_1 \neq -P_1)$ &$(P_2 \neq -P_2)$
$2P = Q = (x_3, y_3)$
$\gamma = (3x_1^2 + a)/2y_1$
$x_3 = \gamma^2 - 2x_1$
$y_3 = \gamma(x_1 - x_3) - y_1$

Equation (2) is scalar multiplication operation in ECC which is explained in ECDLP earlier. Scalar multiplication of a point P occurs k times to account ECC algorithms of smart card systems. Based on this equation, given the prime modulus p, the curve constants from Table 1 a and b and two points in coordinate systems P_1 and P_2 , the problem is to find a scalar $[k]$ that fulfils the equation.

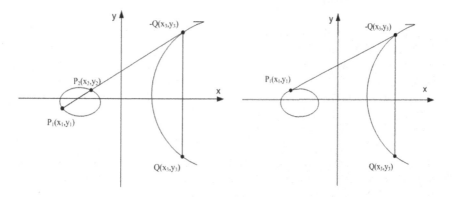

Fig. 6 Geometric explanation of point addition   **Fig. 7** Geometric explanation of point doubling

4 Implementation of ECC

4.1 Design of Simulation

Implementation of ECC over prime field (F_p) is simulated for a smart card system is based on Java card technology. The simulation program was developed using Java Card 2.2.2, Version 1.3 card simulator in Eclipse IDE. Figure 8 illustrates the interactions in the simulation.

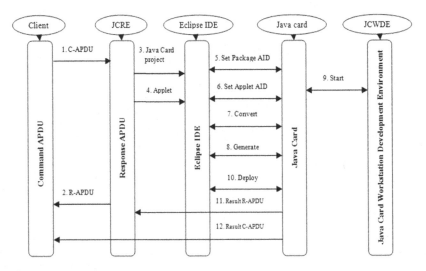

Fig. 8 Java card simulator

4.2 Domain Parameters

Implementation of ECC over prime field (F_p) applies Elliptic Curve Cryptography (ECC) encryption and ECDSA (Elliptic Curve Digital Signature Algorithm) algorithms in smart card algorithms. EC domain parameters of prime field (F_p) are recommended by SEC [21] and NIST [19]. These recommendations are: 160 bit, 224 bit, 256 bit, 384 bit, 384 bit, and 521 bit domain parameters. EC domain parameters over prime field (F_p) are the sextuple:

$$T = (p, a, b, G, n, h) \tag{3}$$

In this sextuple, p specifies prime field (F_p) that two elements $a, b \in F_p$ occur Elliptic Curve $E(F_p)$ and is defined by the equation $y^2 = x^3 + ax + b$: base point $Q = (x_Q, y_Q)$ on $E(F_p)$, a prime n which is the order of Q. An integer h is the cofactor that $h = \#E(F_p)/d$ (#E – Number of points on E). Sextuple of EC domain parameters are specified in ECC, which is represented as an octet string converted using the conventions specified, and they should be determined the following logarithm equations [19, 20].

$$\lceil \log_2 p \rceil \in \{160, 192, 224, 256, 384, 521\} \tag{4}$$

These restrictions are designed to encourage interoperability while providing commonly required security levels. Equation (1.4) of EC domain parameters calls [20]

$$\lceil \log_2 p \rceil = 2t \tag{5}$$

where t is a bit of security. With this equation solving the logarithm problem on the associated elliptic curve would require approximately 2^t operations.

The EC domain parameters over prime field (F_p) are supported at each security level; they are Koblitz curve parameters and verifiably random parameters. However, verifiably random parameters are supplied at export strength and at extremely high strength than Koblitz domain parameters. Mainly, Koblitz domain parameters are suitable EC domain parameters over binary field (F_{2^m}). We implement verifiable random parameters into smart card systems.

4.3 ECC Encryption Algorithm

ECC can be used to encrypt plaintext (message m) through ECC over prime field (F_p) into smart card integrated chips. User randomly chose a base point $G = (x_G, y_G)$ and integer number r which is on the elliptic curve E. These parameters are illustrated in Figure 8. The plain text is coded into an elliptic curve point $P_m = (x_m, y_m)$. Each user selects a private key s and generates his/her public key $P = sG$. Figure 9 further illustrates the ECC encryption scheme, which is to encrypt ECC over prime field (F_p) with the following key lengths in bits:

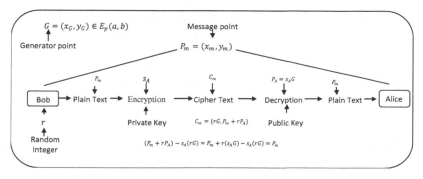

Fig. 9 ECC over prime field (F_p) encryption scheme

160, 192, 224, 256, 384, and 521. These key lengths are compared against RSA encryption key lengths, which are [3] 1024, 1536, 2048, 3072, 7680, and 15360 bits.

Flow chart is illustrated in Figure 10 is a flowchart that illustrates the implementation of the encryption process.

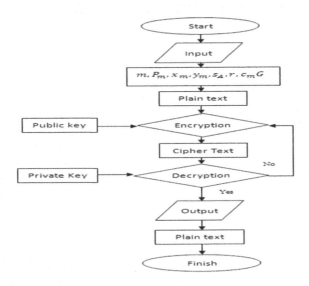

Fig. 10 Flow chart of ECC encryption over prime field (F_p)

4.4 ECDSA Digital Signature Algorithm

ECDSA (Elliptic Curve Digital Signature Algorithm) is the elliptic curve analogue of the Digital Signature Algorithm (DSA) [7, 8, 9]. It is actually a fast digital signature algorithm based on ECC (Elliptic Curve Cryptography) and hash functions (SHA-1). Thus, ECDSA is designed to be existentially unforgivable, even in the presence of an adversary capable of launching chosen-message attacks. We used the same key lengths in bits as in the ECC encryption algorithm: 160, 192, 224, 256, 384, and 521 bits. These are compared with RSA digital signature key lengths, which are 1024, 1536, 2048, 3072, 7680 and 15360 bits, used in SHA-1 hash function.

ECDSA signature scheme is similar to digital signature process that consists of signature generation and signature verification. The following algorithms explain step by step the implementation of the ECDSA digital signature.

Algorithm 1: ECDSA signature generation

Input: Domain parameters $D = (q, FR, s, a, b, P, n, h)$, private key d, message m.

Output: Signature (r, s)

1. Select random or pseudorandom integer k ($k \in [1, n-1]$)
2. Compute scalar multiplication $k * P = (x_1, y_1) or (X_1, Y_1, Z_1)$ and convert x_1 to an integer $\overline{x_1}$
3. $r = \overline{x_1} \mod n$, if $r = 0$ then go to step 1.
4. $k^{-1} \mod n$
5. Hash function $SHA - 1(m)$ which is converted from bit string to integer e
6. Compute $s = k^{-1}(e - dr) \mod n$, if $s = 0$ then go to step 1
7. Return (r, s)

Algorithm 2: ECDSA signature verification

Input: Domain parameters $D = (q, FR, s, a, b, P, n, h)$, Public key Q, message m.

Output: Pass or Fail

1. Verify r and s are integers in the interval $[1, n-1]$
2. Hash function $SHA - 1(m)$ which is converted from bit string to integer e
3. Compute $w = s^{-1} \mod n$
4. Compute $u_1 = ew \mod n$ and $u_2 = rw \mod n$
5. Compute $X = u_1 P + u_2 Q$
6. If $X = O$ (point at infinity) reject the signature. Otherwise, convert the x coordinate x_1 of X to an integer $\overline{x_1}$ and compute $v = x_1 \mod n$
7. Accept the signature $v = r$

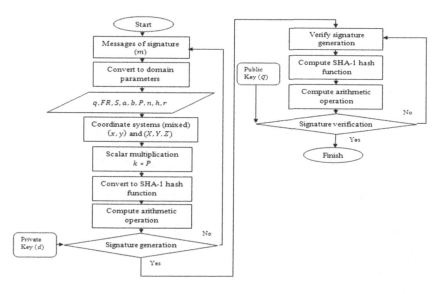

Fig. 11 Flow chart of ECDSA over prime field (F_p)

Figure 11 is a flow chart of ECDSA over prime field (F_p) for the digital signature process is illustrated step by step; this process is based on the above mentioned Algorithm 1 and Algorithm 2.

5 Results

The simulation of encryption and digital signature was performed on a set of smart card data to measure the performance of the algorithms. These are compared against RSA encryption and digital signature algorithms.

Table 2 shows the results of ECC encryption and RSA encryption with different key lengths, where E-time is the encryption time, D-time is the decryption time and total time is the summary of the both times. The results are matched for corresponding key lengths of the different encryption methods as described above. For example, ECC encryption using 160 bit key length is compared against RSA encryption using 1024 bit key length. Here the encryption time of ECC is a slower than RSA encryption time but decryption time is faster than RSA decryption time. Overall the total time for ECC is a smaller than RSA denoting a better performance for ECC method.

Table 2 ECC and RSA encryption simulator results

N	Key size	ECC encryption over prime field (F_p)			Key size	RSA encryption		
		E-time (ms)	D-time (ms)	Total time (ms)		E-time (ms)	D-time (ms)	Total time (ms)
1	160	2.65	1.31	3.96	1024	0.46	8.60	9.06
2	192	3.18	1.57	4.75	1536	0.63	11.83	12.46
3	224	3.71	1.83	5.54	2048	0.85	15.89	16.84
4	256	4.24	2.09	6.33	3072	1.29	24.39	25.68
5	384	6.37	3.14	9.51	7680	3.39	64.41	67.80
6	521	6.64	4.25	12.89	15360	12.71	129.97	142.68

Table 3 shows the results of ECC digital signature (ECDSA) over prime field and RSA digital signature generation and verification with different key lengths, where G-time is the digital signature generation time, V-time is the digital signature verification time and total time is the summation of both times. The results are matched for corresponding key lengths of the different encryption methods as before. For example, ECC digital signature using 160 bit key length is compared against RSA digital signature using 1024 bit key length. Here the digital signature generation time of ECC is lower than RSA but digital signature verification time is higher than RSA. Thus the performance of ECC is better for digital signature generation but the performance of RSA is better for digital signature verification.

Table 3 ECC and RSA digital signature simulation results

N	Key size	ECDSA over prime field (F_p)			Key size	RSA digital signature		
		G-time (ms)	V-time (ms)	Total time (ms)		G-time (ms)	V-time (ms)	Total time (ms)
1	160	36.82	38.05	74.07	1024	75.19	4.67	79.86
2	192	38.22	40.06	78.28	1536	111.57	7.12	118.69
3	224	42.15	45.67	87.82	2048	152.45	9.63	160.47
4	256	55.65	58.33	113.98	3072	228.07	16.64	244.71
5	384	74.83	78.04	152.87	7680	544.92	43.54	588.46
6	521	99.76	103.09	208.85	15360	1137.2	122.15	1259.25

6 Conclusion and Future Work

ECC over prime field cryptography is mathematically proven to be highly secure for encryption and digital signature functions of smart card systems. Initial experiments with a sample data set showed significant improvements in performance compared to RSA methods. Further investigation with a larger set of data is to be conducted to demonstrate the effectiveness of ECC over prime field cryptography and improve the algorithms in providing better performance and security.

References

1. Xiaoyun, W., Yiqun, L.Y., Hongbo, Y.: Finding Collisions in the Full SHA-1, pp. 1–20 (2005)
2. Rankl, W., Effing, W.: Smart card handbook, 3rd edn. John Wiley and Sons Ltd., Springer, Baffins Lane, England (2002)
3. Vincent, O.R., Folorunso, O., Akinde, A.D.: Improving e-payment security using Elliptic Curve Cryptosystem. Electronic Commerce Research 10(1), 27–41 (2010)
4. Diffie, W., Hellman, M.E.: New directions in cryptography. IEEE Trans. IT-22(6), 644–654 (1976)
5. Koblitz, N.: Elliptic Curve Cryptosystem. Mathematics of computation 48, 203–209 (1989)
6. Hendry, M.: Smart Card Security and Applications, 2nd edn. Artech House, Boston (2001)
7. Bart, P.: A survey of recent developments in cryptographic algorithms for smart cards. Computer Networks 51(9), 2223–2233 (2007)
8. Johan, B., Preneel, B., Rijmen, B.: Cryptography on smart cards. Computer Networks 36(4), 423–435 (2001)
9. Vanstone, S.: Elliptic Curve Cryptosystem-The Answer to Strong, Fast Public-Key Cryptography for Securing Constrained Environments. Information Security Technical Report 2(2), 78–87 (1997)
10. Berta, I., Mann, Z.: Implementing Elliptic Curve Cryptography on PC and Smart Card. Periodical Polytechnic Serial El.Eng. 46(2), 47–73 (2002)
11. Johnson, D., Menezes, M., Vanstone, S.: The Elliptic Curve Digital Signature Algorithm (ECDSA), vol. 1, pp. 36–63. Springer, Heidelberg (2001)

12. Hitchcock, Y., Dawson, Y., Clark, A., Montague, P.: Implementing an efficient elliptic curve cryptosystem over GF (p) on a smart card. ANZIAM J. 44, C354–C377 (2003)
13. Mahdavi, R., Saiadian, A.: Efficient scalar multiplications for elliptic curve cryptosystems using mixed coordinates strategy and direct computations. In: Heng, S.-H., Wright, R.N., Goi, B.-M. (eds.) CANS 2010. LNCS, vol. 6467, pp. 184–198. Springer, Heidelberg (2010)
14. Miller, V.S.: Use of elliptic curves in cryptography. In: Williams, H.C. (ed.) CRYPTO 1985. LNCS, vol. 218, pp. 417–426. Springer, Heidelberg (1986)
15. Dabholkar, A., Yow, K.C.: Efficient Implementation of Elliptic Curve Cryptography (ECC) for Personal Digital Assistants (PDAs). Wireless Personal Communications: an International Journal of Verlag New York 29, 233–246 (2004)
16. Hankerson, D., Menezes, A., Vanstone, S.: Guide to Elliptic Curve Cryptography, 1st edn. Springer and Francis Group (2006)
17. Chaves, R., Kuzmanov, G., Sousa, L., Vassiliadis, S.: Cost-Efficient SHA Hardware Accelerators. IEEE Transactions on Very Large Scale Integration (VLSI) Systems 16, 990–1008 (2008)
18. Avanzi, R.M., Cohen, H., Doche, C., Frey, G., Lange, T., Nguyen, K., Vercauteren, F.: Handbook of Elliptic and Hyperelliptic Curve Cryptography. Taylor, Springer (2004)
19. Brown, M., Hankerson, D., López, J., Menezes, A.: Software implementation of the NIST elliptic curves over prime fields. In: Naccache, D. (ed.) CT-RSA 2001. LNCS, vol. 2020, pp. 250–265. Springer, Heidelberg (2001)
20. The Standards for Efficient Cryptography Group (SECG), SEC 1: Elliptic Curve Cryptography, version 2.0 (2000),
    http://www.secg.org/index.php?action=secg,docs_secg

Dynamic Resource Allocation for Improved QoS in WiMAX/WiFi Integration

Md. Golam Rabbani, Joarder Kamruzzaman, Iqbal Gondal,
Iftekhar Ahmad, and Md. Rafiul Hassan

Abstract. Wireless access technology has come a long way in its relatively short but remarkable lifetime, which has so far been led by WiFi technology. WiFi enjoys a high penetration in the market. Most of the electronic gadgets such as laptop, notepad, mobile set, etc., boast the provision of WiFi. Currently most WiFi hotspots are connected to the Internet via wired connections (e.g., Ethernet), and the deployment cost of wired connection is high. On the other hand, since WiMAX can provide a high coverage area and transmission bandwidth, it is very suitable for the backbone networks of WiFi. WiMAX can also provide the better QoS needed for many 4G applications. WiMAX devices, however, are not as common as WiFi devices and it is also expensive to deploy a WiMAX-only infrastructure. An integrated WiMAX/WiFi architecture (using WiMAX as backhaul connection for WiFi) can support 4G applications with QoS assurance and mobility, and provide high-speed broadband services in rural, regional and urban areas while reducing the backhaul cost. WiMAX and WiFi have different MAC mechanisms to handle QoS. WiMAX MAC architecture is connection-oriented providing the platform for strong QoS control. In contrast, WiFi MAC is not connection-oriented, hence can provide only best effort services. Delivering improved QoS in an integrated WiMAX/WiFi architecture poses a serious technological challenge. The paper depicts a converged architecture of WiMAX and WiFi, and then proposes an adaptive resource distribution model for the access points. The resource distribution model ultimately allocates

Md. Golam Rabbani · Joarder Kamruzzaman · Iqbal Gondal
Faculty of IT, Monash University, VIC, Australia
e-mail: golam.rabbani@monash.edu, g.rabbani@csebuet.org,
    joarder.kamruzzaman@monash.edu, iqbal.gondal@monash.edu

Iftekhar Ahmad
School of Engineering, Edith Cowan University, WA, Australia
e-mail: i.ahmad@ecu.edu.au

Md. Rafiul Hassan
Department of Information and Computer Science, KFUPM, Dhahran, KSA
e-mail: mrhassan@kfupm.edu.sa

R. Lee (Ed.): Software Eng., Artificial Intelligence, NPD 2011, SCI 368, pp. 141–156.
springerlink.com   © Springer-Verlag Berlin Heidelberg 2011

more time slots to those connections that need more instantaneous resources to meet
QoS requirements. A dynamic splitting technique is also presented that divides the
total transmission period into downlink and uplink transmission by taking the minimum data rate requirements of the connections into account. This ultimately improves the utilization of the available resources, and the QoS of the connections.
Simulation results show that the proposed schemes significantly outperform the
other existing resource sharing schemes, in terms of maintaining QoS of different
traffic classes in an integrated WiMAX/WiFi architecture.

1 Introduction

WiFi has contributed greatly to the growth in high-speed Internet services through
its efficient wireless access technology. In the past several years, a number of munic-
ipalities and local communities around the world have taken the initiative to deploy
WiFi systems in outdoor settings in order to provide broadband access to cities and
metrozones as well as rural and underdeveloped areas. Over 97% of notebooks today
come with WiFi as a standard feature, and an increasing number of handhelds and
Consumer Electronics (CE) devices are adding WiFi capabilities [1]. Because of the
wide popularity of the WiFi technology, major vendors of handheld devices, such
as Google and Apple, include WiFi compatible radio interfaces in their most popular mobile devices including Android, iPhone, iPad and iPod touch, etc. Despite
its wide popularity, WiFi is not free from drawbacks. Small range, limited mobility,
and the inability to provide strict quality of service (QoS) for different classes of
services are the biggest drawbacks of WiFi. Moreover, WiFi needs wired backhaul
network for Internet connectivity, which is cost-intensive, especially installation in
areas that lack infrastructure.

WiMAX, the Worldwide Interoperability for Microwave Access, is the much-
anticipated wireless technology aiming to provide business and consumer wireless
broadband access on the scale of the Metropolitan Area Network. The distinctive
features of WiMAX include very high peak data rate, scalable bandwidth and data
rate, adaptive modulation and coding, flexible and dynamic per user resource allocation, and support for advanced antenna techniques. All of these features enable
WiMAX to support improved QoS, high mobility, and robust security. With its large
coverage area and high transmission rate, WiMAX can serve as a backbone for
WiFi hotspots to be connected to the Internet. With mobility support, WiMAX has
become the key technology for users to access high-speed Internet on buses, express
trains [2], ships, etc. As for example, the Utah Transit Authority (UTA) in the USA,
launched the first "free to all passengers" on board wireless broadband network on
a U.S. railroad, which uses WiMAX for the backhaul network [2].

Though WiMAX offers some clear benefits over WiFi, it faces some key challenges for its widespread deployment: i) WiMAX service will cost end users 50-
100% more compared to WiFi services, making it even more difficult to bridge
the so-called digital divide between rich and poor citizens, ii) widespread consumer adoption of WiMAX devices is far lower than WiFi devices, iii) WiMAX

communication using licensed spectrum depends on the cooperation of the monopoly owner of the local spectrum, and iv) supporting clients across a large coverage area with a single WiMAX BS requires an enormous frequency spectrum.

Therefore, considering the widespread penetration of WiFi into the consumer market and long-range capability of WiMAX, the convergence of WiMAX and WiFi using WiMAX as the backbone for WiFi access points (APs), offers an excellent opportunity for the service providers to provide wireless broadband Internet services in the rural and regional as well as urban areas. As the handling of QoS by MAC Mechanisms is different in WiFi and WiMAX, providing QoS in the integrated network poses a great challenge.

To solve the above mentioned problem, our contributions are following: i) presents a resource distribution scheme at the WiFi AP, and ii) proposed a dynamic splitting technique that divides the total transmission period into downlink and uplink transmission.

2 Related Works

WLANs are connected to the Internet and/or other WLANs by a wired backhaul network. Therefore, high deployment costs are associated with the backhaul network of WiFi, particularly in the remote rural or suburban areas with low population densities or new settlement areas lacking wired infrastructure. In contrast, WiMAX is a highly promising technology for the provision of high-speed wireless broadband services. The promise of WiMAX to provide backhaul support for WLANs has enormous commercial appeal. The convergence of these technologies in seamless transmission of data ensuring end user QoS requirements is a major challenge.

Wireless networks like WLAN, HiperLAN/2 are able to coexist [3], i.e., they can operate in the same time and place without having harmful interference on each-other using dynamic frequency selection and controlling transmit power. The coexistence of WiMAX with WiFi in shared frequency bands is discussed in [4]. Here, Berlemann *et al.* proposed coexistence between both standards without exchanging any data between both standards. DCF(Distributed Coordination Function) is used as the MAC(Medium Access Control) in WiFi nodes for enabling the coexistence of a single 802.16 system with multiple 802.11 APs and stations (STAs). The major drawback of this solution is that a significant portion of the available transmission time is wasted which causes reduced spectral efficiency. As, the MACs of 802.11 and 802.16 access the channel alternately, it will be a challenging task for maintaining the QoS of the delay sensitive applications.

Another example of coexistence of IEEE 802.16 and 802.11(e) in the shared frequency bands is depicted in [5]. The common frame structure is also used here. They considered a central coordinating device which combines the BS of 802.16 with the HC(Hybrid Coordinator) of the 802.11e, which is referred as BSHC (Base Station Hybrid Coordinator). The BSHC can operate in both an 802.16 and 802.11 modes. The BSHC uses OFDM-based physical layer which both the standards support. For 802.16 SS, the BSHC is a normal BS while for 802.11e stations, it is an ordinary

HC. The BSHC has full control over the channel and over all 802.16 SSs and 802.11 STAs. The drawback of this coexistence is that the BSHC needs complete control over the radio resources of the co-located 802.11e and 802.16 wireless networks. Also, the QoS requirements of different types of applications fails to be addressed.

Other than the coexistence in shared frequency band approach, hierarchical approach is also recently proposed in literature. In [6], Hui-Tang *et al.* proposed a hierarchical architecture of the integration of WiMAX and WiFi. The WiFi AP will communicate with the WiMAX BSs as well as WiFi nodes. So, they must have two interfaces to support both WiFi and WiMAX MAC and PHY layers and have been termed as W^2-AP. A two-level bandwidth allocation scheme was used in their architecture. The WiFi nodes request for bandwidth to W^2-AP and the W^2-AP accumulates all requests and send to WiMAX BS. The W^2-AP then schedules among the WLAN nodes. However, they did not give any description about the scheduling where a good scheduling algorithm is required to provide guaranteed QoS. In [7], Gakhar *et al.* proposed a QoS architecture for the interworking of WiMAX and WiFi to map the QoS requirements of an applications originating at the WiFi network to the WiMAX network. They did not provide the mechanisms required to ensure QoS like bandwidth allocation, scheduling and admission control. Integration of WiFi with WiMAX mesh network is also proposed in [8].

Utility function can be used for the resource allocation problem [9, 10, 11, 12]. In [9], Kelly *et al.* use utility function for the analysis of stability and fairness of rate control algorithm. In [10, 11], utility function is employed to build a bridge between physical layer and media access control (MAC) layer. In [12], Shi *et al.* also uses utility for scheduling purposes, but the author did not consider the priority of the connections in the utility function.

WiMAX defines five QoS service classes [13] to provide support for a wide variety of applications which are: 1) Unsolicited Grant Services (UGS) - this service class provides a fixed periodic bandwidth allocation for the application. There is no need to send any other requests after the connection setup. It is designed to support applications that require fixed-size data packets transmission at a constant bit rate (CBR) such as E1/T1 circuit emulation, VoIP without silent suppression; 2) Real-time Polling Service (rtPS) - this is designed to support variable bit rate real-time traffic such as MPEG compressed video; 3) Extended Real-time Polling Service (ertPS) - this service class is designed to support real-time applications, such as VoIP with silence suppression, that have variable data rates, but require guaranteed bit rate and delay. This service class is similar to UGS in that the BS allocates the maximum sustained rate during active mode, but unlike UGS, no bandwidth is allocated during silent period; 4) Non-real-time Polling Service (nrtPS) - this service class is designed to support non-real-time VBR traffic which does not require delay guarantee like File Transfer Protocol (FTP); 5) Best Effort Service (BE) - this service class is designed to support data streams that do not require a minimum service-level guarantee. Most of the data traffic belongs to this service class. WiMAX can deal with various applications using these service classes. On the other hand, WiFi provides best efforts service to the various types of applications. It is a big challenge to provide the WiFi end users WiMAX-like QoS where WiMAX is used as backhaul network for the WiFi APs. In the following, we first describe the network architecture that we are going to use for convergence which is followed by the adaptive scheduling algorithm to support class-wise QoS in the later section where we have proposed a utility function for the resource allocation algorithm in WiFi APs.

3 The Hierarchical Network Architecture

A single IEEE 802.16 BS serves two types of connections: standalone Subscriber Stations (SSs) and WLAN APs. Fig. 1 shows a BS serving a SS and a WLAN AP. For standalone SS, the connection is dedicated to a single user while the WLAN connection is shared among the WLAN nodes. For standalone SS, the uplink and downlink data is directly transferred through the connection with the SS. In case of WLAN nodes, the data is transferred through WLAN AP. The WLAN AP has dual transceiver which can work using both 802.11 and 802.16 interfaces.

The WLAN nodes individually request bandwidth to the WLAN AP. The WLAN AP first maps the traffic requests to WiMAX QoS service classes and then sends bandwidth request for each of the service classes to the WiMAX BS. The IEEE 802.11 MAC layer provides two types of mechanisms to access the channel - DCF and PCF (Point Coordination Function) [14]. DCF is a distributed access control mechanism which is a randomized protocol employing CSMA/CA (Carrier Sense Multiple Access with Collision Avoidance) and provides a best effort service only. PCF is a centralized MAC algorithm to provide contention-free service to support for delay sensitive applications. As PCF is a centralized MAC algorithm, it can offer better QoS to the connections than DCF. We propose to use PCF in WLAN AP to offer better QoS. For the purpose of providing prioritized medium access, we propose a scheduling-based mechanism on managing transmissions in PCF. Scheduling in WiMAX BS is a well-investigated topic. So, we will concentrate on scheduling in WiFi AP.

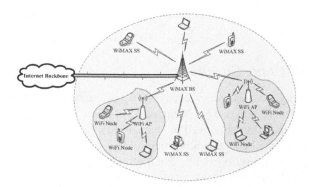

Fig. 1 Hierarchical network model of the integration of WiMAX and WiFi

4 QoS Compliance Adaptive Scheduling

The PCF provides contention-free frame transfer where the stations do not have to contend for access. In PCF, the point coordinator (AP) polls the WiFi nodes in the polling list following round-robin (RR) scheduling. Since the basic RR-based scheduling in 802.11 PCF access mechanism does not consider priority at all, we propose a utility-based scheduling in PCF so that higher priority traffics are given appropriate consideration in polling. Two scheduling algorithms are proposed here - one for downlink traffic and another for uplink traffic which is due to the different behavior of uplink and downlink traffic from the end-users point of view. The contention free period (CFP) is divided into "downlink transmission period" and "uplink transmission period". In the utility function (Eq. (2)), Eq. (3) is used as the priority index for downlink traffic and Eq. (4) is used for uplink traffic. Using the utility function, we formulate an optimization problem in Eq. (5), the solution of which will yield the polling schedule of the WiFi nodes. We assume that the WiFi nodes can request bandwidth as they require. From that bandwidth request, the 'QoS Class' and 'minimum rate' of the connection are obtained and the connections for uplink traffic can be scheduled. Eq. (4) is used as the priority index for uplink traffic because the APs don't have enough information about delay and queue length of uplink packets.

Let us assume K WiFi connections to be supported by the WiFi AP. The contention-free period of PCF is divided into N time slots. Our scheduling scheme aims to distribute these N slots among the K connections so that the priorities of the connections are maintained as well as the minimum rates are served. Let $x_{k,n}$ be an element of matrix X with K rows and N columns. Each element $x_{k,n}$ is defined as:

$$x_{k,n} = \begin{cases} 1, & n\text{-th slot is assigned to } k\text{-th connection} \\ 0, & n\text{-th slot is not assigned to } k\text{-th connection} \end{cases}$$

Similarly let $R_{k,n}$ be the number of bits transmitted by the k-th connection if n-th slot is assigned to k-th connection. The total number of bits transmitted by the k-th connection in one cycle is

$$r_k = \sum_{n=1}^{N} (R_{k,n} \times x_{k,n}) \tag{1}$$

4.1 Adaptive Utility Function

We design a utility function which attributed by a number of network parameters including applications data rate requirement, QoS class, urgency to meet delay requirement and fairness factor defined by the network provider. Utility function can be used to evaluate the benefit of allocating a time slot to certain connection. The purpose of using the utility function is that it ultimately allocates more time slots to the connections that need more instantaneous resources to meet QoS requirement, e.g., a connection whose delay limit is approaching or whose buffer is approaching full will have its utility function adjusted as per priority class and required data rate. The utility function is adaptive to the connections' and network's current state.

The utility function has to be an increasing function and the marginal utility function has to be a decreasing function. Because, with the increasing utility function, the utility-based resource allocation can exploit the multi-user diversity and the user will be given timeslot according to their needs. The decreasing marginal utility function can prevent the resource allocation algorithm from favoring excessively the top priority users and maintain fairness of the resource allocation. Here, we choose logarithmic function as the form of the utility function which is an increasing function and has decreasing marginal utility function. The term $F(F > 0)$ in the utility function, defined as fairness factor, is used to adjust the tradeoff between efficiency and fairness of resource allocation. The bigger the value of F is, the more fair the resource allocation is to the users and the lower the spectral efficiency becomes. This type of utility function was used in literature [12] but they did not consider the priority of the connections. The Priority Index, PI in the utility function, represents the urgency of the user's allocation of resources. If the delay, currently experienced by a packet, is near to maximum delay limit, the PI value increases for the connection related to that packet. Similarly, if the queue length of a connection reaches the maximum queue length, the PI value increases for that connection. The connection with low priority and low minimum rate has low PI value. Thus the term PI in the formulation is adaptive to the changing traffic load and network condition.

With K Connections IDs (CIDs) and N time slots, let us define the following parameters.

Maximum Delay Limit, d_{max}
Delay Limit at time t, d_k
Maximum Queue Length, q_{max}
Queue Length at time t, q_t
Maximum Rate, $r_{k,max}$
Minimum Rate, $r_{k,min}$
QoS Class, Q_k
Fairness Index, F
Utility Function, $U_k(r_k)$

We define the utility function as follows,

$$U_k(r_k) = \begin{cases} \log_{(1+F)}(1 + F \times PI_k \times \dfrac{r_k}{r_{k,max}}), & r_k \leq r_{k,max} \\ \log_{(1+F)}(1 + F \times PI_k), & r_k > r_{k,max} \end{cases} \tag{2}$$

Where PI_k for uplink and downlink traffic is defined as

$$PI_k = \frac{Q_k \times r_{k,min}}{(q_{max} - q_t) \times (d_{max} - d_t)} \text{ (For downlink traffic)} \tag{3}$$

$$PI_k = Q_k \times r_{k,min} \text{ (For uplink traffic)} \tag{4}$$

4.2 Optimization

To allocate the time slots to the connections maintaining the QoS, the summation of utility functions of the connections needs to be maximized. To formulate an optimization problem based on the utility function for the time slots allocation, let K connections can use N slots in such a way that the summation of the utility functions of all the connections is maximized subject to the conditions that each slot will be allocated to one connection only and the minimum data rate of each of the connections is ensured. The optimization problem can be formulated as follows:

$$\max_{x_{k,n}} \sum_k U_k(r_k) \tag{5}$$

$$\text{Subject to} \begin{cases} x_{k,n} \in \{1,0\}, & \forall n, \forall k \\ \sum_{k=1}^{K} x_{k,n} = 1, & \forall n \\ r_k \geq r_{k,min}, & \forall k \end{cases} \tag{6}$$

One slot will be assigned to only one connection which is ensured by the 1st and 2nd constraint in Eq. (6). The minimum data rate of each connection is ensured by the 3rd constraint in Eq. (6).

The solution to the optimization problem involves both continuous variable PI_k and binary variable $x_{k,n}$. For K connections and N time slots, there are K^N possible ways of time slot allocation to the connections. The process of finding an optimal solution, therefore, may not be always robust. Considering the practical applications, we propose a heuristic algorithm in the next section which produces a near-optimal solution and at the same time computationally inexpensive.

4.3 Heuristic Algorithm

In our heuristic algorithm, the connections are sorted according to their QoS classes. We allocate slots to the connections so that the minimum rate of each of the connections is satisfied starting from the highest priority QoS class connections to the lowest priority. For the rest of the slots, for each of the connections, we calculate the difference of the total utility function before and after allocating that slot to a connection. The connection with the highest difference is allocated that slot. It is worthwhile to compare the computational complexity of our heuristic algorithm with the exhaustive method. In the heuristic algorithm (detailed steps presented in Algorithm 1), in phase (1), we need $O(N)$ computation and in phase (2) we need $O(KN)$ computation. Therefore, overall time complexity of our heuristic algorithm is $O(KN)$ which is far less than the time complexity of the exhaustive method $O(K^N)$.

Algorithm 1. Heuristic Allocation Algorithm

1: PHASE 1: *Allocate timeslot for minimum rate requirement*
2: $allocation\_index \leftarrow 0$
3: Sort the connections according to QoS classes
4: Let the sorted connections are $1, 2 ... K$
5: **for** $k \leftarrow 1$ to K **do**
6:     Allocate minimum number of slot p to Connection k so that $r_k = r_{k,min}$
7:     **for** $i \leftarrow 1$ to P **do**
8:         $allocation\_index \leftarrow allocation\_index + 1$
9:         $x_{k,allocation\_index} \leftarrow 1$
10:         **if** $allocation\_index = N$ **then**
11:             **break**
12:         **end if**
13:     **end for**
14:     **update** r_k
15:     **if** $allocation\_index = N$ **then**
16:         **break**
17:     **end if**
18: **end for**
19: PHASE 2: *Allocate the rest of the slots*
20: **for** $n \leftarrow allocation\_index$ to N **do**
21:     **for** $k \leftarrow 1$ to K **do**
22:         $utility1 = \sum_k U_k(r_k)$
23:         $x(k, n) \leftarrow 1$
24:         **update** r_k
25:         $utility2 = \sum_k U_k(r_k)$
26:         $x(k, n) \leftarrow 0$
27:         **update** r_k
28:         $diff\_utility_{k,n} \leftarrow utility2 - utility1$
29:     **end for**
30:     find $k* = \arg \max_k(diff\_utility_{k,n})$
31:     $x_{k*,n} \leftarrow 1$
32:     **update** r_k*
33: **end for**

---

5 Dynamic Allocation of CFP of PCF in Downlink and Uplink Transmission

In the PCF mechanism, there is no control over the ratio of downlink and uplink data. In standard PCF, the WiFi nodes are polled by the WiFi AP according to a round-robin algorithm. The WiFi AP polls a WiFi node if it has downlink data to send to that node. When the WiFi node receives the downlink data, it will send uplink data if it has any. Therefore, there is no control at the AP over the ratio of downlink and uplink data. We implemented our resource distribution algorithm at WiFi AP for both downlink and uplink data. Here, we concentrate on the dynamic allocation of the CFP of PCF in downlink and uplink data, as shown in Fig. 2.

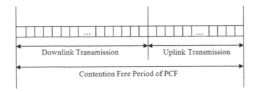

Fig. 2 Allocation of CFP of PCF into Downlink and Uplink Transmission

Data transmission is symmetrical for VoIP-like applications where the same amount of data is sent both ways. However, in case of Internet connections or broadcast data (e.g., streaming video), it is likely that more data will be sent in the downlink transmission period than in the uplink transmission period. Hence, equal split ratio between downlink and uplink data will degrade the system performance, while unequal but fixed split ratio will still lead to poor utilization of available resources. We propose a mechanism of splitting the CFP in downlink and uplink transmission periods as shown in Algorithm 2. The CFP is divided according to the minimum number of slots required to maintain the minimum data rate for the downlink and uplink connections, respectively.

Algorithm 2. Dynamic Allocation of CFP

1: $N \leftarrow$ Total number of slots
2: Find the minimum number of slots, $N\_downlink\_req$ required for all downlink connections to meet minimum data rate requirement
3: Find the minimum number of slots, N_uplink_req required for all uplink connections to meet minimum data rate requirement
4: $N\_downlink\_alloc \leftarrow \frac{N \times N\_downlink\_req}{N\_downlink\_req + N\_uplink\_req}$
5: $N\_uplink\_alloc \leftarrow N - N\_downlink\_alloc$
6: Use $N\_downlink\_alloc$ and N_uplink_alloc as the total number of slots in the scheduling algorithm for downlink and uplink data respectively

---

6 Results

In this section, we present some illustrative numerical results which reveal the performance of the proposed scheduling algorithm in PCF of WiFi in the hierarchical architecture of the integration of WiMAX and WiFi. We used a single-channel infrastructure WLAN in our simulation using custom simulator *atisim* developed by M.A.R. Siddique in his PhD thesis [15]. Though we considered different combinations of number of connections and time slots, due to space limitation, here we show only the results of experiment considering 30 connections and 40 PCF time slots for both downlink and uplink connections. QoS classes '3' refers to UGS, '2' refers to rtPS/ertPS/nrtPS and '1' refers to BE in these results.

To the best knowledge of the authors, there is no other relevant work in literature that deals with the QoS in WiMAX/WiFi integration architecture at the WiFi AP level. Hence, we compare our method against PCF used in WiFi AP where round-robin method is used. Round-robin method assigns time slots to each connection in equal portions and in circular order, handling all connections without priority. Fig. 3(a) and Fig. 3(c) show the average class-wise throughput of our proposed scheme and its round-robin counterpart for downlink connections. Fig. 5(a) and Fig. 5(c) present the same data for uplink connections. As evident in Fig. 3(a) and Fig. 5(a), our method yields higher throughput to the higher priority connections whereas round-robin method used in PCF, as expected, yields same throughput to all the connections irrespective of their priority. This is because we consider the priority of the connections in our utility function which ultimately affects time slot allocation. So, higher priority connections get higher throughput. Traditional PCF does not consider priority among the connections. Fig. 3(b) show the time averaged throughput for each connection in each method for downlink connections and Fig. 5(b) present the same data for uplink connections. Here we see the same scenario that the higher priority connection are provided improved throughput and QoS.

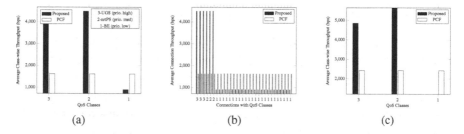

(a)   (b)   (c)

Fig. 3 Comparison of performance between our proposed method and standard PCF for **downlink data** (a) Average class-wise throughput for K = 30 (b) Average connection throughput for K = 30 (c) Average class-wise throughput for K = 20

We also investigated queue occupancy in both methods for downlink connections as it has significant impacts on delay. In Fig. 4(a) ~ 4(c), we compare the leftover space in the queue for the UGS, rtPS/ertPS/nrtPS and BE individual connections respectively between our proposed method and PCF. Results show that for UGS connections, the leftover space in queue is higher in our method than PCF. In case of rtPS/nrtPS connections, again our method provides more space in queue than PCF. In case of BE connections, PCF provides more space than our method while our method provides substantial space in the queue for the BE connections. Our method considers the priority of the connections while distributing the time slots to the connections. The leftover space in the queue is also considered in the priority index. By doing so, our method serves higher priority class more while still serving the lower priority connections.

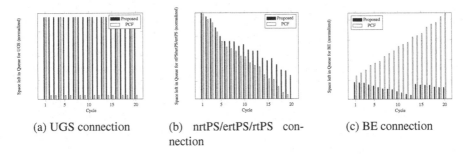

(a) UGS connection   (b) nrtPS/ertPS/rtPS con-   (c) BE connection
nection

Fig. 4 Comparison of performance between our method and standard PCF in terms of space left in the queue

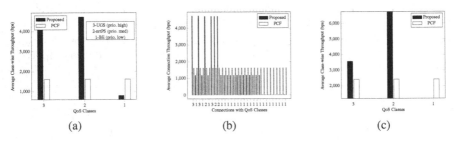

(a)   (b)   (c)

Fig. 5 Comparison of performance between our proposed method and standard PCF for **uplink data** (a) Average class-wise throughput for K = 30 (b) Average connection throughput for K = 30 (c) Average class-wise throughput for K = 20

Table 1 QoS (minimum rate requirement) violation rate (%) for **downlink connections**

QoS Class (Averaged)	QoS (minimum rate req.) violation rate in our algorithm		QoS (minimum rate req.) violation rate in PCF	
	K = 15	K = 30	K = 15	K = 30
UGS	0.58	0	12.78	98.68
rtPS/ertPS/nrtPS	0	0.79	12.78	98.68
BE	0	0	0	0

We also calculated the normalized number of times when the minimum rate could not be delivered to the connections in both methods as shown in Table 1 for downlink connections and Table 2 for uplink connections. It shows that with 15 connections and 40 time slots, the minimum rate requirement of the connections were almost always satisfied in our method. In case of PCF, the minimum rate requirement of the UGS and rtPS/nrtPS is not satisfied in about 12.78% of time for both downlink and uplink connections. Our method performs even better when there is scarcity of resource (30 connections and 40 time slots) also shown in Table 1 and Table 2 for downlink and uplink connections, respectively.

Table 2 QoS (minimum rate requirement) violation rate (%) for **uplink connections**

QoS Class (Averaged)	QoS (minimum rate req.) violation rate in our algorithm		QoS (minimum rate req.) violation rate in PCF	
	K = 15	K = 30	K = 15	K = 30
UGS	0.60	0.60	12.78	98.68
rtPS/ertPS/nrtPS	0	0.40	8.52	98.68
BE	0	0	0	0

We implemented the dynamic allocation of CFP of PCF in downlink and uplink transmission by MAC of AP in *atisim*. We present the data for both downlink and uplink connections using a fixed splitting method and our proposed dynamic splitting algorithm. For the fixed splitting method, we considered the ratio of downlink and uplink transmission to be 3:1. Although we used different combinations of the

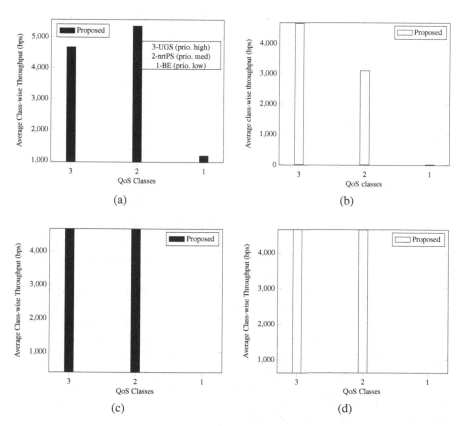

Fig. 6 Average class-wise throughput for (a) **downlink** connections using **fixed** splitting ratio (b) **uplink** connections using **fixed** splitting ratio (c) **downlink** connections using **dynamic** splitting ratio (d) **uplink** connections using **dynamic** splitting ratio

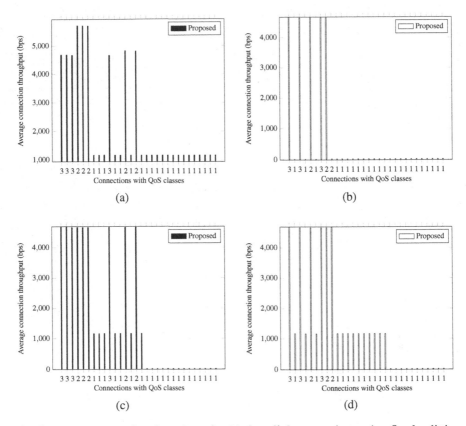

Fig. 7 Average connection throughput for (a) **downlink** connections using **fixed** splitting ratio (b) **uplink** connections using **fixed** splitting ratio (c) **downlink** connections using **dynamic** splitting ratio (d) **uplink** connections using **dynamic** splitting ratio

total number of connections and time slots, the results for 30 connections with the total number of slots (80) are presented here.

Fig. 6(a) and Fig. 6(b) present the average class-wise throughput for downlink and uplink connections, respectively, using fixed splitting ratio 3:1. Fig. 6(c) and Fig. 6(d) show the average class-wise throughput for downlink and uplink connections, respectively, using our proposed dynamic splitting algorithm. Here, we see that the average class-wise throughputs of 'rtPS/nrtPS/ertPS' and 'BE' uplink connections using fixed splitting are smaller when compared with connections using our proposed dynamic allocation algorithm. Moreover, the average class-wise throughputs of 'rtPS/nrtPS/ertPS' and 'BE' downlink connections attained more time slots than were required, in the case of fixed splitting. This is because the fixed splitting method does not consider the nature of the traffic, and always gives the same number of time slots to uplink and downlink connections. Our method, however, takes the minimum rate requirements of the connections into consideration and allocates the time slots to downlink and uplink connections accordingly. So, the network resources are more efficiently utilized by means of our method. Fig. 7(a) and Fig. 7(b) provide the average connection throughput for each of the downlink and uplink connections, respectively, using a fixed splitting ratio, whereas, Fig. 7(c) and Fig. 7(d) show the average connection throughput for each of the downlink and uplink connections, respectively, using our proposed dynamic allocation method. These figures present the same scenario more clearly; namely, that the uplink connections attain less time slots in the case of a fixed splitting ratio. In Fig. 7(b), we see that the average throughput of the 9th connection (which belongs to QoS class '2') is very low and cannot fulfill its minimum data rate requirement. The average throughputs of other QoS class '2' connections are also low compared with the results of our proposed method.

7 Conclusion

In this paper, we have discussed a converged WiMAX/WiFi network architecture where we use PCF in the WiFi APs. We proposed a QoS-aware adaptive scheduling algorithm in PCF at the WiFi APs to allocate the time slots among the WiFi nodes. To provide QoS support to connections, we defined a utility function that contains priority index, a measure that takes QoS class, minimum data rate, queue status and packet delay into account. The scheduling algorithm maximizes the total utility of all connections to allocate the time slots over a cycle. We also proposed a dynamic mechanism for dividing the contention free period of PCF into downlink and uplink transmission period. An algorithm was formulated in consideration of the minimum rate requirements of the connections in allocating time slots to downlink and uplink transmissions. The schemes, we proposed in this paper, were supported by simulation results. Our results show that our method significantly outperforms the traditional round-robin algorithm used in PCF in terms of meeting QoS requirements. We also showed that fixed or equal ratio of splitting leads to poor utilization of available resources and degradation of QoS requirements, whereas dynamic splitting performs significantly better. Future work will focus on developing an efficient admission control mechanism so that the utilization will be maximized as well as the minimum rate and latency requirements of the flow can be guaranteed without hampering other flows considering the priority of the new flow and existing flows.

References

1. WiMAX and WiFi Together: Deployment Models and User Scenarios,
   http://www.motorola.com/staticfiles/Business/Solutions/
   Industry%20Solutions/Service%20Providers/
   Wireless%20Operators/Wireless%20Broadband/wi4%20WiMAX/
   _Document/StaticFile/WiMAX_and_WiFI_Together_Deployment_
   Models_and_User_Scenarios.pdf
   (last accessed July 8, 2010)

2. (January 2010),
   http://www.dailywireless.org/2010/01/15/
   free-wifi-on-acela-express-trains/
3. Mangold, S., Habetha, J., Choi, S., Ngo, C.: Co-existence and Interworking of IEEE
   802.11a and ETSI BRAN HiperLAN/2 in MultiHop Scenarios. In: Proc. of IEEE 3rd
   Workshop in Wireless Local Area Networks, Boston, USA, September 27-28 (2001)
4. Berlemann, L., Hoymann, C., Hiertz, G.R., Walke, B.: Unlicensed Operation of IEEE
   802.16: Coexistence with 802.11(A) in Shared Frequency Bands. In: 2006 IEEE 17th
   International Symposium on Personal, Indoor and Mobile Radio Communications,
   September 11-14, pp. 1–5 (2006)
5. Berlemann, L., Hoymann, C., Hiertz, G.R., Mangold, S.: Coexistence and Interworking of IEEE 802.16 and IEEE 802.11(e). In: Proc. of IEEE 63rd Vehicular Technology
   Conference, VTC 2006-Spring, Melbourne, Australia, May 7-10 (2006)
6. Hui-Tang, L., et al.: An Integrated WiMAX/WiFi Architecture with QoS Consistency
   over Broadband Wireless Networks. In: 6th IEEE Consumer Communications and Networking Conference, CCNC 2009 (2009)
7. Gakhar, K., Gravey, A., Leroy, A.: IROISE: a new QoS architecture for IEEE 802.16 and
   IEEE 802.11e interworking. In: 2nd International Conference on Broadband Networks,
   BroadNets 2005, October 7, pp. 607–612 (2005)
8. Niyato, D., Hossain, E.: Integration of IEEE 802.11 WLANs with IEEE 802.16-based
   multihop infrastructure mesh/relay networks: A game-theoretic approach to radio resource management. IEEE Network 21(3), 6–14 (2007)
9. Kelly, F.P., Maullooa, A.K., Tan, D.K.H.: Rate control for communication networks:
   shadow prices, proportional fairness and stability. Journal of the Operational Research
   Society 49(3), 237–252 (1998)
10. Song, G.C., Li, Y.: Cross-Layer Optimization for OFDM Wireless Networks-Part I:
    Theoretical Framework. IEEE Transactions On Wireless Communication 4(2), 614–624
    (2005)
11. Song, G.C., Li, Y.: Cross-Layer Optimization for OFDM Wireless Networks-Part II:
    Algorithm Development. IEEE Transactions On Wireless Communication 4(2), 625–634
    (2005)
12. Shi, J., Hu, A.: Maximum Utility-Based Resource Allocation Algorithm in the IEEE
    802.16 OFDMA System. In: IEEE International Conference on Communications, ICC
    2008, May 19-23, pp. 311–316 (2008)
13. So-In, C., Jain, R., Tamimi, A.-K.: Scheduling in IEEE 802.16e mobile WiMAX net-
    works: key issues and a survey. IEEE Journal on Selected Areas in Communica-
    tions 27(2), 156–171 (2009)
14. Siddique, M.A.R., Kamruzzaman, J.: VoIP Capacity over PCF with Imperfect Channel.
    In: IEEE Global Telecommunications Conference, GLOBECOM 2009, November 30-
    December 4, pp. 1–6 (2009)
15. Siddique, M.A.R.: Voice Quality Framework for VoIP over WLANs, PhD thesis, GSIT,
    Monash University, Australia (2011)

Automated Classification of Human Daily Activities in Ambulatory Environment

Yuchuan Wu, Ronghua Chen, and Mary F.H. She

Abstract. This paper presents a human daily activity classification approach based on the sensory data collected from a single tri-axial accelerometer worn on waist belt. The classification algorithm was realized to distinguish 6 different activities including standing, jumping, sitting-down, walking, running and falling through three major steps: wavelet transformation, Principle Component Analysis (PCA)-based dimensionality reduction and followed by implementing a radial basis function (RBF) kernel Support Vector Machine (SVM) classifier. Two trials were conducted to evaluate different aspects of the classification scheme. In the first trial, the classifier was trained and evaluated by using a dataset of 420 samples collected from seven subjects by using a k-fold cross-validation method. The parameters σ and c of the RBF kernel were optimized through automatic searching in terms of yielding the highest recognition accuracy and robustness. In the second trial, the generation capability of the classifier was also validated by using the dataset collected from six new subjects. The average classification rates of 95% and 93% are obtained in trials 1 and 2, respectively. The results in trial 2 show the system is also good at classifying activity signals of new subjects. It can be concluded that the collective effects of the usage of single accelerometer sensing, the setting of the accelerometer placement and efficient classifier would make this wearable sensing system more realistic and more comfortable to be implemented for long-term human activity monitoring and classification in ambulatory environment, therefore, more acceptable by users.

1 Introduction

Automatic recognition and classification of human daily activities have attracted substantial interests in many applications, such as health care or rehabilitation

Yuchuan Wu · Ronghua Chen · Mary F.H. She
Institute for Technology Research and Innovation
Deakin University, 3216 Victoria, Australia

Yuchuan Wu
School of Mechanical Engineering and Automation
Wuhan Textile University, 430073 Wuhan, China

R. Lee (Ed.): Software Eng., Artificial Intelligence, NPD 2011, SCI 368, pp. 157–168.
springerlink.com   © Springer-Verlag Berlin Heidelberg 2011

[1-2], sports [3] or aged care [4]. Various sensing technologies and classification schemes have been developed to tackle this scientific issue in the past decades. For instance, Kubo et al. [5] used a micro-wave Doppler sensor to monitor human motion by measuring the distance change between the target and the sensor; Pressure sensors underneath a floor (GRF) were also used to classify human movements based on the vertical component measurements of the ground reaction force. Many others employed computer-vision technologies for human motion analysis [6-7]. However, these technologies are impractical in the context of classifying human daily activities in ambulatory environment.

Generally, there are several other issues need to be addressed to promote the use of long-term activity monitoring system in ambulatory environment. These include ease of use, comfort and safety of wearing, discretion and the ability to perform daily activity unimpeded. Any system which impedes those functions is most likely to be rejected by users.

Recently, appreciable research efforts have been devoted toward the automated monitoring and classifying human or animal physical activities from wearable sensor data [8-10]. Accelerometers are currently among the most widely used body-worn sensors for long-term activity monitoring and classification in free-living environment due to its small size, light weight, cheap price as well as low power consumption [11-13]. Because of that, the sensory data from accelerometers have been used to detect falls, evaluate human mobility or goats' grazing behavior. Theses sensors are usually attached onto various placements, including chest, waist, legs, arms and back, etc. For the sake of safety and comfort of wearing, a single accelerometer sensor is more preferable to be attached onto the waist belt.

Although this wearable sensing technology offers an optimal platform for monitoring daily activity patterns if appropriately implemented, effective feature extraction and robust classification algorithms are also required for representing the signals in a more informative and compact format and subsequently interpreting them in the context of different activities. It is highly desirable but still challenging to design a robust human activity classification approach based on the sensory data from a single accelerometer for the sake of safety and comfort of the wearers.

While discrete wavelet transform (DWT) has gained widespread acceptance as a useful feature extraction technology in single processing, many classification techniques have been developed in the past decades, such as Bayesian decision making, K-nearest neighbor (k-NN) algorithm, Rule-based algorithm [14], Least-squares method (LSM), Dynamic time warping [15], Artificial neural networks (ANN), Hidden Markov model (HMM) [16], etc. Support vector machine (SVM) is an important machine learning technique which was originally proposed in the early 1980s for the recognition and classification of objects, voices, and text or handwritten characters. The advantage of the SVM classifier is that SVM is able to pre-process and represent feature vectors in a higher-dimensional space where they can become linearly separable, if the feature vectors in their original feature space are not linearly separable [17].

In this work, we proposed an approach to human daily physical activities classification based on the sensory data from a single tri-axial accelerometer. The accelerometer was attached to human subjects' waist belt to capture three channels of gravitation signals. Then, a set of feature vectors were extracted from the sensory data through wavelet transform, followed by dimension reduction using Principle Component Analysis (PCA). Finally, a RBF kernel Support Vector Machine (SVM) classifier was employed to make decision about the class of the wearers' physical activities. Specifically, the optimal parameter σ and c of the RBF kernel were obtained by searching automatically. The classifier was validated on the acceleration data collected from thirteen subjects performing six different physical activities.

2 Experimental Setup

Apparatus

For data acquisition, a sensing unit containing a single tri-axial accelerometer sensor (MMA7260, Freescale) and a microcontroller (C8051F020, Silicon Laboratories Inc.) with a 12-bit analog-to-digital convertor (ADC) was designed. The tri-axial accelerometer measures the acceleration along three orthogonal axes, with a sampling rate of 50Hz which is much higher than human activity frequency (less than 20Hz [18]) and a full-scaled affection fixed at +4g (where g represents acceleration due to gravity: 9.81 m/s^2). The original data by microcontroller were sent to a laptop computer with USB interfaces, at a communication rate 9600bps. Equation (1) gives the relationship between input voltage signal (V) and acceleration (G) for the tri-axial accelerometer as g-range is +4g.

$$G = 3.33 \times V - 5.5 \qquad (1)$$

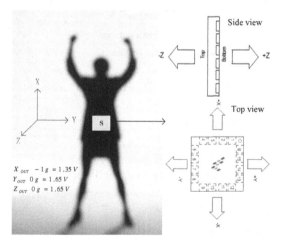

Fig. 1 Reference directions and position of tri-axial accelerometer sensor (S) mounted on the waist belt

According to the findings in the ergonomics studies [19], the tri-axial accelerometer was attached to the subjects' waist belt, given the consideration that the
waist belt would be more comfortable and safer compartment for embedding an
accelerometer.

The reference directions and laid position of tri-axial accelerometer are shown
in Fig. 1.

Subjects

Thirteen healthy volunteers (11 males and 2 females) were recruited from the Institute for Technology and Research Innovation, Deakin University, Australia.
The subjects varied in age from 26 and 50 years, in weight from 51kg to 75kg and
in height from 158cm to 179cm. All of them were in good health, and none of
them had a special deformity in their bodies and limbs.

Acceleration Data

Acceleration data were collected from 13 volunteers performing six activities, including falling, jumping, running, sitting-down, standing, walking and falling.
Seven subjects (1# to 7#), whose data were used to train can cross-validate the
classifier, were asked to repeat every action for ten times, while the remaining six
subjects (8# to 13#), whose data were used to test the generalization capability of
the system, were asked to perform each activity for twice.

The activity labels and instructions are shown in table 1.

Table 1 Activity labels and instructions

Label	Activity	Instruction
1	falling	fall forward down
2	jumping	jump at origin
3	running	run forward
4	sit-down	sit down on chair about 5 seconds
5	standing	remain standing straight
6	walking	walk forward

In this way, a total of 492 kinematic datasets were recorded from 13 volunteers,
containing 420 training datasets from subjects 1# to 7# and 72 testing datasets
from subjects 8# to 13#.

3 Recognition Algorithms

In this study, the building blocks of the activity identification and verification system are based on stages used in typical pattern recognition systems. First, raw da-

ta were pre-processed by segmenting interest signals from original kinematic data, i.e., the action signal data. Second. To make activity classification at next stage more effective and efficient, the discrete wavelet transform (DWT) was used to present acceleration data in a more informative and compact format through decomposing the signal into several components [20]. Then, principal component analysis (PCA) was adopted to reduce the dimensionality of the feature vector [21], which is aimed to make the recognition and classification module more efficient and realistic for real-time applications. Finally, SVM classifier was employed for activity classification [22-23]. The algorithms were developed in MATLAB [TM] environment.

Activity Patterns

Fig. 2 shows the four original signal patterns of six different activities from the same volunteer (1#).

From Fig.2, it can be observed that jumping and falling activities have clearly distinguishable characteristics compared to the other four activities. There is substantial different between walking and running on the signal amplitude, though the patterns of the signals look similar. Moreover, the signals of walking and running by different subjects are hardly discriminated either in amplitude or in frequency.

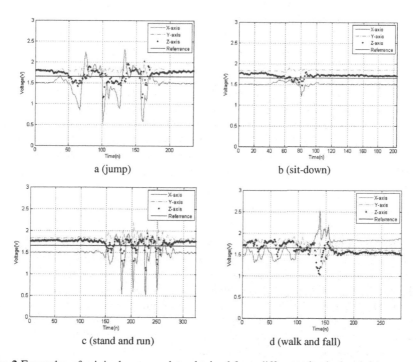

a (jump)   b (sit-down)

c (stand and run)   d (walk and fall)

Fig. 2 Examples of original sensory data obtained from different physical activities

Feature Extraction Based on DWT

It is necessary to prepare training samples and testing samples to establish and evaluate the classifier. In this work, signals of interest associated with each activity from original time series of acceleration data were segmented into individual datasets.

In this work, feature extraction was accomplished by applying discrete wavelet transform (DWT) to present the acceleration signals corresponding to activities in a more informative and compact format. DWT is a widely accepted technology to decompose a signal into different frequency bands. The signal with n subsamples, $x(t) = \{x(1),...,x(n)\}$ is decomposed with a low-pass filter by using a wavelet function $\varphi(t)$ to extract the approximation signal $A(t)$ and with a high-pass filter by using the scaling function $\psi(t)$ to extract the detail signal $D(t)$. Only the approximation signal $A(t)$ is used for the next level of decomposition. The wavelet decomposition into approximation and detailed signals are defined as follows [24]:

$$A_i(t) = \sum_{k=-\infty}^{\infty} A_{i-1}(k)\phi_i(t-k) \tag{2}$$

$$D_i(t) = \sum_{k=-\infty}^{\infty} A_{i-1}(k)\psi_i(t-k) \tag{3}$$

The reconstruction of the wavelet coefficient is defined as

$$x(t) = \sum_{k=-\infty}^{\infty} A_L(k)\phi_i(t-k) + \sum_{i=1}^{L} \sum_{k=-\infty}^{\infty} D_{i-1}(k)\psi_i'(t-k) \tag{4}$$

For DWT, the values of Ai(t) are further down sampled at next level.

In order to extract feature vectors, statistics over the set of the wavelet coefficients was used. The extracted wavelet coefficients provide a compact representation that shows the energy distribution of the human activity signal in time and frequency. The approximate coefficients (Ca) were taken as extracted feature vectors.

Dimension Reduction by PCA

To reduce the complexity of classifier and the number of samples needed for building a classification system, principle component analysis was used to further reduce the size of the feature vectors. As such, each feature vector was normalized and projected to the most discriminative data array. In other words, the original data set is rotated to the direction of maximum variance, where correlated high-dimensional data can be presented in a low-dimensional uncorrelated feature space and the data variance is preserved as much as possible with a small number of principal components, i.e., the largest eigenvectors [25].

In Fig.3, the energy level of the high dimensional data retained by different numbers of the largest coefficients obtained by PCA was displayed. It is observed that the first 48 eigenvectors are sufficient enough to retain 92% energy of the data. Therefore, their eigenvalues will serve as inputs into classifier at the next stage.

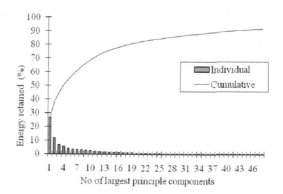

Fig. 3 Percentage of energy retained by largest principle components

SVM Classifier

Unlike other supervised pattern recognition methods which aim only to minimize the classification error, SVM simultaneously minimizes the classification error and maximizes the geometric margin. It can automatically adjust its capacity according to the scale of a specific problem by maximizing the width of the classification margin [23]. This is achieved by picking the hyperplane so that the distance from the hyperplane to the nearest data point is maximized.

Given data $\{x1,...,xNt\}$ with their labels $\{Li=1, -1\}$, the hyperplane takes the form wTx−b=0, where w is a normal vector perpendicular to the hyperplane. Moreover, all data satisfy the following constraint:

$$w^T x_i - b \geq +1 \quad \text{for} \quad L_i = +1 \tag{5}$$

$$w^T x_i - b \leq -1 \quad \text{for} \quad L_i = -1 \tag{6}$$

This constraint can be written as:

$$L_i(w^T x_i - b) \geq 1 \tag{7}$$

Now, let us consider the hyperplanes when the two equalities in (5) and (6) hold. These points lie on the hyperplane H_1: wTxi−b=1; and the points lie on the hyperplane H_2: wTxi−b=-1. By using some geometry knowledge, the distance between H_1 and H_2 is $2/\|w\|$. The problem now is to maximize $2/\|w\|$, or equivalently, minimize $\|w\|$ subject to the constraint (7). More precisely,

$$\min\|W\|^2 \text{ s.t. } L_i(w^T x_i - b) \geq 1, \quad I = 1, \cdots, N_t \tag{8}$$

In this paper, radial basis function (RBF) was chosen as kernel function of SVM, where an important step was to find a set of optimal parameters, width of RBF kernel (σ) and the regularization parameter (c), to obtain high classification accuracy. Unlike the manual comparison approach used in previous works [26], the optimal parameters σ and c of RBF kernel SVM was obtained by a grid search approach in this classification algorithm. The underneath principle is to produce a series of values between N and M (i.e. -10 to +10) for parameter c and σ, respectively, followed by automatic searching for the optimal parameter set σ and c which yields the highest classification accuracy.

Fig. 4 describes the relationship between classification accuracy and parameter set of σ and c. The abscissa and ordinate were demoted by logarithm to the base 2. It can be observed that the best σ and c for this example were 0.25 and 4, respectively.

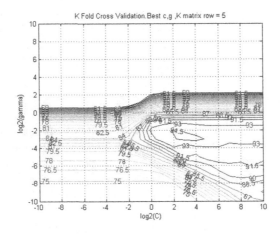

Fig. 4 Relationship of classification accuracy with parameters σ and c

4   Results and Discussion

The classification performance was examined in two experiments. The results obtained during testing in the two trials are shown in Tables 2 to 3, respectively.

Trial 1

Trial 1 was designated to obtain the optimal parameters σ and c which achieved the best classification performance in the training. In this trial, 5-fold cross validation was performed to make sure that the training is un-biased. Specifically, in each process, the dataset was divided into five equal subsets, among which, four subsets were used to train the classifier, and the left one subset was used as cross validation samples. This process is then repeated five times (the folds), with each of the five subsets used exactly once as the validation data. The verification performance of the classifier was assessed by the mean accuracy of 5 processes.

The 420 data collected from subjects 1# to 7# were randomly partitioned into training dataset (336 samples) and cross-validation dataset with 84 samples left. Table 2 shows the recognition results of cross-validation samples.

Table 2 Recognition results in trial 1

Action	Pro 1	Pro 2	Pro 3	Pro 4	Pro 5	Acc1 (%)
fall	14	14	14	13	14	98.60
jump	13	14	9	14	13	90
run	13	13	14	14	12	94.30
sit-down	14	13	14	14	14	98.60
stand	14	13	14	14	14	98.60
walk	12	14	12	11	14	90
acc2(%)	95.20	96.40	91.70	95.20	96.40	95

Pro: process;   acc: accuracy;
acc1: average accuracy for five processes;
acc2: average accuracy for single process

As shown in Table 2, the average recognition accuracy for six activities was 95%.

Trial 2

To further evaluate generalization capability of the classifier, the second trial was carried out and the results are shown in Table 3.

In trial 2, 72 samples from six subjects (8# to 13#), who were unknown to the classifier, were tested.

Table 3 Recognition results in trial 2

Action	Fall	Jump	Run	Sit-down	Stand	Walk	Acc
fall	11	1	0	0	0	0	92%
jump	0	11	1	0	0	0	92%
run	0	1	10	0	0	1	83%
sit-down	0	1	0	12	0	0	100%
stand	0	0	0	0	12	0	100%
walk	0	0	1	0	0	11	92%

(Note: each row represents the test results).

It can be observed that average recognition accuracy of 93% was achieved in this trial.

Discussion

This classification approach has demonstrated that the SVM algorithm can distinguish six human activities based on their corresponding kinematic data collected from the accelerometer sensor. However, the classifier does not perform equally well on different activities. The activities including falling, sitting-down and standing can be recognized very well, while some activities such as jumping, running and walking are relatively prone to misclassification.

It can be observed that the sensory data collected from different people performing the same activities such as running and walking show different kinematic characteristics in the sensory data. One subject's running characteristic could be similar to another's walking action in terms of frequency and/or amplitude. Because of this, the classifier is easily confused by running and walking signals.

Due to the placement of tri-axial accelerometer sensor on the waist belt (refer to Fig.1), the y-axial signals haven't been basically changed over the time during sampling the six activities. Therefore, it is acceptable that the signals from y-axial can be removed in future experiment. As a positive effect, it would consequently improve the recognition accuracy further.

5 Conclusion

In this preliminary study, we have presented a novel human daily activity classification approach based on the sensory data collected from a single tri-axial accelerometer worn on the waist belt.

The classification algorithm was realized to distinguish 6 different activities including standing, jumping, sitting-down, walking, running and falling through three major steps: discrete wavelet transformation (DWT), Principle Component Analysis (PCA)-based dimensionality reduction and followed by implementing a radial basis function (RBF) kernel Support Vector Machine (SVM) classifier . The parameters σ and c of the RBF kernel were optimized through automatic searching in terms of yielding the highest recognition accuracy and robustness.

492 sets of signals obtained from 13 subjects were obtained to evaluate the algorithm in two trials. 420 datasets were used to train and cross-validate the classifier in the first trial, while 72 datasets were used to test its generalization capability in the second trial. An overall mean recognition accuracy of 95% and 93% were achieved in trials 1 and trial 2, respectively, by this activity classifier. The results obtained show the system is not only able to learn the activity signals from the subjects whose data were involved in training but also has good generalization capability for the data from new subjects. Moreover, compared to other placements, the position of the accelerometer sensor on the waist belt offers easier implementation and causes less discomfort to the wearers in real-time applications. It is also very beneficial to recognize human daily activities using as few sensors as possible for applications in health care, rehabilitation, aged care or sport, etc. Therefore, the collective effects of the usage of single accelerometer sensing, the setting of the accelerometer placement and efficient classifier would make this

wearable sensing system more realistic and more comfortable to be implemented for long-term human activity monitoring and classification in ambulatory environment, therefore, more acceptable by users.

It can be concluded that the approach to the recognition and classification of multiple human daily activities based on the sensory data of a single tri-axial accelerometer is promising and will be further explored in the near future.

References

1. Doukas, C., Maglogiannis, I.: Advanced classification and rules-based evaluation of motion, visual and biosignal data for patient fall incident detection. International Journal on Artificial Intelligence Tools 19(2), 175–191 (2010)
2. Salarian, A., et al.: Ambulatory monitoring of physical activities in patients with Parkinson's disease. IEEE Transactions on Biomedical Engineering 54(12), 2296–2299 (2007)
3. Bonomi, A.G., et al.: Detection of type, duration, and intensity of physical activity using an accelerometer. Medicine and Science in Sports and Exercise 41(9), 1770–1777 (2009)
4. Schwartz, A.V., et al.: Diabetes-related complications, glycemic control, and falls in older adults. Diabetes Care 31(3), 391–396 (2008)
5. Kubo, H., Mori, T., Sato, T.: Detection and measurement of human motion and respiration with microwave Doppler sensor. In: 23rd IEEE/RSJ 2010 International Conference on Intelligent Robots and Systems (2010)
6. Wang, L.A., Suter, D.: Recognizing human activities from silhouettes: Motion subspace and factorial discriminative graphical model. In: 2007 IEEE Conference on Computer Vision and Pattern Recognition, vol. 1-8, pp. 2518–2525 (2007)
7. Zhang, J.G., Gong, S.G.: Action categorization by structural probabilistic latent semantic analysis. Computer Vision and Image Understanding 114(8), 857–864 (2010)
8. Bang, S., et al.: Toward Real Time Detection of the Basic Living Activity in Home Using a Wearable Sensor and Smart Home Sensors. In: 2008 30th Annual International Conference of the Ieee Engineering in Medicine and Biology Society, vol. 1-8, pp. 5200–5203. IEEE, New York (2008)
9. Patel, S., et al.: Using Wearable Sensors to Monitor Physical Activities of Patients with COPD: A Comparison of Classifier Performance. In: Lo, B., Mitcheson, P. (eds.) Proceedings of the Sixth International Workshop on Wearable and Implantable Body Sensor Networks, pp. 234–239. IEEE Computer Soc., Los Alamitos (2009)
10. Wu, J.K., et al.: Real-time physical activity classification and tracking using wearble sensors. In: 2007 6th International Conference on Information, Communications & Signal Processing, vol. 1-4, pp. 1697–1702. IEEE, New York (2007)
11. Denkinger, M.D., et al.: Accelerometer-based physical activity in a large observational cohort - Study protocol and design of the activity and function of the elderly in Ulm (ActiFE Ulm) study. BMC Geriatrics 10 (2010)
12. Kang, D.W., et al.: Real-time elderly activity monitoring system based on a tri-axial accelerometer. Disability and Rehabilitation: Assistive Technology 5(4), 247–253 (2010)
13. Lai, C.F., et al.: Detection of cognitive injured body region using multiple triaxial accelerometers for elderly falling. IEEE Sensors Journal 11(3), 763–770 (2011)

14. Altun, K., Barshan, B., Tunçel, O.: Comparative study on classifying human activities with miniature inertial and magnetic sensors. Pattern Recognition 43(10), 3605–3620 (2010)
15. Pogorelc, B., Gams, M.: Discovery of gait anomalies from motion sensor data. In: International Conference on Tools with Artificial Intelligence, ICTAI (2010)
16. Rabiner, L.R.: Tutorial on hidden Markov models and selected applications in speech recognition. Proceedings of the IEEE 77(2), 257–286 (1989)
17. Maulik, U., Chakraborty, D.: A self-trained ensemble with semisupervised SVM: An application to pixel classification of remote sensing imagery. Pattern Recognition 44(3), 615–623 (2011)
18. Antonsson, E.K., Mann, R.W.: The Frequency Content of Gait. Journal of Biomechanics 18(1), 39–47 (1985)
19. Kangas, M., et al.: Comparison of low-complexity fall detection algorithms for body attached accelerometers. Gait and Posture 28(2), 285–291 (2008)
20. Chen, Y.J., Oraintara, S., Amaratunga, K.S.: Structurally regular biorthogonal filter banks: Theory, designs, and applications (2004)
21. Shahbudin, S., et al.: Analysis of PCA based feature vectors for SVM posture classification. In: 6th International Colloquium on Signal Processing and Its Applications (CSPA 2010), Melaka (2010)
22. Das, K., Giesbrecht, B., Eckstein, M.P.: Predicting variations of perceptual performance across individuals from neural activity using pattern classifiers. NeuroImage 51(4), 1425–1437 (2010)
23. Lau, H.Y., Tong, K.Y., Zhu, H.: Support vector machine for classification of walking conditions using miniature kinematic sensors. Medical and Biological Engineering and Computing 46(6), 563–573 (2008)
24. Kandaswamy, A., et al.: Neural classification of lung sounds using wavelet coefficients. Computers in Biology and Medicine 34(6), 523–537 (2004)
25. Wu, J., Wang, J.: PCA-based SVM for automatic recognition of gait patterns. Journal of Applied Biomechanics 24(1), 83–87 (2008)
26. Fukuchi, R.K., et al.: Support vector machines for detecting age-related changes in running kinematics. Journal of Biomechanics 44(3), 540–542 (2011)

A Reinforcement Learning Approach with Spline-Fit Object Tracking for AIBO Robot's High Level Decision Making

Subhasis Mukherjee, Shamsul Huda, and John Yearwood

Abstract. Robocup is a popular test bed for AI programs around the world. Robo-soccer is one of the two major parts of Robocup, in which AIBO entertainment robots take part in the middle sized soccer event. The three key challenges that robots need to face in this event are manoeuvrability, image recognition and decision making skills. This paper focuses on the decision making problem in Robosoccer–The goal keeper problem. We investigate whether reinforcement learning (RL) as a form of semi-supervised learning can effectively contribute to the goal keeper's decision making process when penalty shot and two attacker problem are considered. Currently, the decision making process in Robosoccer is carried out using rule-base system. RL also is used for quadruped locomotion and navigation purpose in Robosoccer using AIBO. Moreover the ball distance is being calculated using IR sensors available at the nose of the robot. In this paper, we propose a reinforcement learning based approach that uses a dynamic state-action mapping using back propagation of reward and Q-learning along with spline fit (QLSF) for the final choice of high level functions in order to save the goal. The novelty of our approach is that the agent learns while playing and can take independent decision which overcomes the limitations of rule-base system due to fixed and limited predefined decision rules. The spline fit method used with the nose camera was also able to find out the location and the ball distance more accurately compare to the IR sensors. The noise source and near and far sensor dilemma problem with IR sensor was neutralized using the proposed spline fit method. Performance of the proposed method has been verified against the bench mark data set made with Upenn'03 code logic and a base line experiment with IR sensors. It was found that the efficiency of our QLSF approach in goalkeeping was better than the rule based approach in conjunction with the IR sensors. The QLSF develops a semi-supervised learning process over the rule-base system's input-output mapping process, given in the Upenn'03 code.

Keywords: Reinforcement Learning, semi-supervised, Robocup, Aperius, Robosoccer, IR sensros, camera.

Subhasis Mukherjee · Shamsul Huda · John Yearwood
Centre for Informatics and Applied optimization
GSITMS, University of Ballarat, Victoria, Australia
e-mail: smukherjee@academic.mit.edu.au, s.huda@ballarat.edu.au,
       j.yearwood@ballarat.edu.au

R. Lee (Ed.): Software Eng., Artificial Intelligence, NPD 2011, SCI 368, pp. 169–183.
springerlink.com   © Springer-Verlag Berlin Heidelberg 2011

1 Introduction

Robocup is an international event which is divided in two different sections namely Robosoccer and Rescue league [1]. Both the fields are currently popular test beds for AI programs around the world and the AIBO robot takes part into the middle size soccer event there. Usually a team of four robots play soccer against the opponent team. Some other robots that take part in other events are QRIO [2], ASIMO [3], AIBO [4], ACTROID [5] and so on. The AIBO robots were considered as the medium of the experiment in this case. The last released model ERS-7 has enough processing power to compute relatively simple program and capable of multitasking in real time mode as it runs on Aperius [6] real time operating system. The three key challenges that robots need to face in this event are manoeuvrability, image recognition and decision making skills. We focus on the decision making skills mixed with ball distance and measurement problem. In this paper we consider the goal keeper problem for decision making in Robosoccer. The first and basic problem is goalkeeping against one attacker. The attacker shots the ball from different positions towards the goal. The second problem is an extension of the first one. The knowledge base achieved by the goalkeeper against one attacker was used to save the goal against two attackers as well. The first attacker passes the ball to its mate while it takes shot towards the goal using the flying pass. Both the problems were accompanied by a spline fit method for monitoring the ball's location and distance using nose camera. Due to the noisy data achieved with conventional IR sensors a new measurement method was introduced.

In this paper we investigate whether reinforcement learning (RL) as a form of semi-supervised learning can effectively contribute to the goal keeper's decision making process when a penalty shot is taken. At the same time a new ball distance method was developed using the ball finder method[19] and nose camera. The same decision making process was introduced for a two attacker problem to test its usefulness later on. In the literature [18] review it was found that RL was used to determine the low level functions such as quadruped locomotion and navigation purpose using AIBO. The decision making process was carried out using predefined rule-based systems which requires hand coded technique and shows some inherent drawbacks due to the fixed rule-base nature of the process. We propose here a Reinforcement Learning (RL) based approach combined with Q-Learning with spline fit (QLSF) for AIBO to train it for choosing the best action in a given situation in order to perform goalkeeping. The RL (Q-learning) can produce a dynamic mapping to the intercepted state and a set of actions and helps learn the AIBO to act independently during the game so that it can find an optimum solution taking the complete problem space into account. Thus the proposed QLSF procedure is able to overcome the drawbacks of out of Vocabulary (OOV) condition in existing static rule-base technique. Previous IR sensor detection method for incoming ball was not efficient enough because of the other objects in the vicinity that incorporated noise in the data. As a result we proposed a method using the 2D vision of the nose camera. Also the camera was pointing at the middle point of the ball using an inbuilt function. As a result the head-pan angle was readily used as a direction of the ball. As a result the agent was able to locate the ball before choosing an action.

The rest of the paper is organized as follows.

- The second section briefs the literature review of the Robosoccer and related works.
- Theories and a touch of the working principle of the proposed algorithm are discussed in the third section.
- Experiments result and discussion section elaborates the design of the experiment. A comparative study over the obtained result and the benchmark is also described properly over there.
- Finally the conclusion reveals the contribution of the paper and future framework.

2 Related Work

The history of robotics started long before the introduction of Robosoccer tournament. The first electronics autonomous robot was discovered by the middle of the 19th century [7]. According to the history of Artificial Intelligence system, if-else logic was used to create an expert system to unlock the mystery of an expert answering machine system. Although preliminary system showed little success, it was realized later on that a real life system will take much more than the prototype [17]. Soon scientists found out that the incorporation of human expertise needs the same amount of processing power it takes the human brain to run. However, beside the idea of making an intelligent computer program the idea of making a system learn evolved as well. Basically a learning process could be of three types: supervised semi-supervised and completely unsupervised. According to the society of robotics, the semi-supervised technique or a grey-box approach is one of the leading approaches currently using possible pattern matching system [8]. In the year 1999 the LRP [9] team wins Robocup due to the superiority of their walking program over the others. The program gave a stable view of the filed compare to the others and so achieved the victory. A significant ball controlling technique was introduced in Robocup 2000 by the winning team UNSW [10]. They defended their title in 2001 Robocup [11] using the new model of AIBO ERS-210 [4]. A lot of changes were introduced during this year. Carnegie Melon University won the 2002 title using the strategy control technique. rUnswift [12] team of UNSW won the robocup 2003 again using a very effective and complex locomotion process. The program enabled their robots with precise movement solution. However the process was mainly based on static programming method. Later in 2005[13] RL and fuzzy logic both were combined to make the AIBO learn how to walk and tune its speed and other manoeuvring technique.

The ball detection and finding technique in the game was determined with the help of both the nose camera and IR sensor together. Usually the camera is used to locate the balls position and the sensors was to detect an approximate value its distance. However, the gap existed in sole use of the learning process for making decision in Robosoccer and also the elimination of the IR sensor due to the noise in its data generated from other objects in the vicinity. The next section discusses a methodology involving decision making using RL in goalkeeping and efficient and sole use of the camera to locate ball distance and direction.

3 Proposed Methodology: Q Learning with Spline Fit in (QLSF) Goalkeeping in Robosoccer

The proposed methodology investigates the application of reinforcement learning in Robosoccer and use of nose camera to find out the location and the distance of the ball. The main focus is on applying a specific RL technique in decision making for goalkeeping. The aim here is to use an RL technique with the AIBO for it to learn the best action in a given situation to perform goalkeeping. Reinforcement learning is a dynamic situation-to-action mapping that helps the actor to gain comfort or reward through the process. An action is chosen in order to response to a situation. Based on the outcome of the action taken, a reward or punishment is given. Finally a converged reward matrix announces the end of the training. Exploration versus exploitation dilemma is one of them that have to be countered during the RL process design. The choice of this factor greatly affects the convergence time of the reward matrix.

The existing ball distance determination technique using IR is too noisy because of the other objects in the vicinity. These objects could be either an obstacle or a player (friend and/or foe) and any obstacle that reflects the IR back. In addition to the two different IR sensors are available in AIBO to measure obstacles exist within 20cm and above that. As a result when the ball lies at the transition range (approximately at 20cm range) it is critical to choose either the near or far sensor. Before, describing the detail methodology, some the key term has been introduced here to clarify the detail procedure.

3.1 Related Theories

Policy →A policy is a way to map from state to action. It could be used to define the balance between exploration versus exploitation. Soft policies ensure enough room for the exploration while exploiting the rest of the time.

On-policy learning →It is said to be on-policy learning if an agent starts learning with a particular policy and finds out the state action values within the scope of the policy. SARSA [14] is one of the on-policy approaches exist.

Off-policy learning →In contrast with the on-policy learning, the agent starts updating the table with a strict greedy policy in case of the off-policy learning. However, a predefined degree of freedom also provides enough room for exploration. Finally this method comes up with a new policy at the end of the training. Due to this feature, this method was incorporated to tackle the research question under the Q-learning method.

Different action selection policies →There different action selection policies exist such as ε-greedy and Softmax. A ε-greedy policy [15] always tries to exploit the obtained knowledge as it is called ε-greedy. However, a considerable degree of freedom is always given to ensure the exploitation. The same policy is chosen to train the robot during the training process.

Value Function →A value function is a state action pair function that estimates the return to any state after an action is taken. The following expression gives an essence of its numerical form.

V^{Π} (S,A) →The value of a state S under policy Π. To be precise the expected return when starting in S and following Π thereafter.

State x action table →State x action table is used as the data base of the learning agent to store its knowledge during the training process.

Temporal Difference →Temporal difference (TD) method is used to estimate value function discussed above. Once the final reward was received, the complete path taken to reach the final state could be traced back using the temporal difference equation as stated below. An estimate of the final reward is calculated at each state x action value updated at every step on the way.

$$V(S_t) \leftarrow V(S_t) + \alpha * [R_{t+1} - \gamma * V(S_{t+1}) - V(S_t)] \qquad (1)$$

S_t = State visited at time t, R_{final} = Reward received at the end
R_{t+1} = Reward received at time t+1, α = Constant parameter
γ = Discount Facto

Spline method →In mathematics, a spline is a special function defined piecewise by either polynomial or liner equation. A two point linear equation form is used in this regard.

$$y-y_1 = m* (x-x_1) \qquad (2)$$

m = gradient, x,y,x_1,y_1 = Two different point pair

In this paper, the AIBO was given a chance to learn goalkeeping itself. Few penalty shots from different fixed points within the penalty area were given during the learning process. There after a relatively critical situation involving two different players were introduced during the testing period. The agent acted up to the mark as the reinforcement learning method maps the basic experience to cope up with the complex situation. The ball distance measurement was first taken using IR sensor and it was found that the noisy data was interfering with the decision making process and at a point the agent intercepted few situations erroneously.

At first, a simulated maze learning scenario [16] was considered as a starting point for the choice of a proper method to solve the research problem. The reason of the consideration is hidden in the similarity of the given scenario and that of the Robosoccer. The simulated scenario is based on a maze with a goal set for the agent. The agent's task is to find the shortest path to the goal from any of the existing locations. According to the finding the Q-learning [15] yielded a converged matrix within 3000 training epochs using 85% greedy policy. Whereas the SARSA was never be able to produce a converged matrix due to the fact that it only considers the reward generated by the last state, action pair. In contrast with that, Q-learning considers the best of the all available state, action pair from the last step. As a result the final reward matrix becomes more and more optimum each time the agent completes a training epoch. Due to noiseless environment of the simulated process and the goalkeeping area the value of α was considered as 1. The similarity of the goalkeeping process to that of the maze learning is given below.

- The similarity of the size of the state x action table between both the cases
- Significantly less convergence time of the proposed method compared to the SARSA due to the similarity of the environment in terms of involvement of noise.

Proposed Q-learning Algorithm
A step-wise procedural approach based on Q-learning algorithm is displayed in the following figure

```
Initialize Q(S,A) Arbitrarily
repeat
        Initialize S
        Choose A from A(Sₜ) using random selection policy
        For each step do
                Take action A, observe S and A´
                Choose A´ and S´ with random policy
                Update table using Q(S,A) ← Q(S,A) + α * [R
                + γ maxₐ*  Q(S´,A´) - Q(S,A)]
                S ← S´
                A ← A´
                A* ε  A//A
        End for
Until terminal S reached
Return Q(S, A)
end
```

3.2 A Brief Description of the Working Principle

Let us consider that the goal keeping agent starts learning using the Q-learning equation as displayed in the algorithm. A random selection policy is used to choose the action as the state-to-action mapping is not available at the beginning. According to the Q-learning theory it will not receive a result until it reaches a final state. Once it receives a terminal state, a reward is received using the equation

$$Q(S,A) \leftarrow Q(S,A) + \alpha * [R + \gamma \max_A* Q(S^{'},A') - Q(S,A)] \qquad (3)$$

However, because of the terminal state the $\gamma \max_A* Q(S^{'},A')$ parameter will have zero value and the full amount of the available reward value will be stored to the reward matrix. Furthermore during another training epoch sometimes the agent reaches a state where a previously succeeded state-action pair can be considered as the next one using 85% ε-greedy policy. On the other hand the spline method is applied on a table of values taken with the ball.ratio [21] parameter while the the nose camera pointed at the middle point of the ball. The ball.ratio yields the percentage value of the area occupied by the ball with the 2-Dimensional image using the nose camera. The value decreases with the distance increases. A value of the ball.ratio was collected at 33 points with 3cm spacing between two points. A spline fit method was applied to the data obtained from this experiment. Detailed and mathematical discussion of both the methods is described in the next section during the discussion of the experiment design.

3.3 Choice of the Robot

The AIBO [4] was chosen as the research platform for the research. It is a quadruped robot with a camera at its nose and few IR sensors around the nose and chest.

Table 1

CPU	64-bit RISK Processor
CPU Clock Speed	576 MHz
RAM	64 MB
Programming media	Memory Stick^{TM}
Operating temperature	10°C to 60°C
Operating humidity	10–80%
Built in sensors	Temperature Sensor, IR distance sensor, Acceleration sensor, Touch sensor, vibration sensor, Pressure sensor
Movable parts	Three parts in head module, Three parts in each of the four legs
Power Consumption	Approximately 9W (Standard operation in autonomous mode)
Operating time	1.5 Hour (In standard operation)
LCD Display on charger	Time date, volume, Battery condition
Operating system	Aperius
Weight	Approximately 1:6 Kg including battery and Memory stick
Dimension	180mm high and 18mm in diameter.

The reason of using the robot is hidden in its mobile mechanism. Usually a wheeled robot can move much faster and in a stable manner than a legged robot. However, an uneven terrain is the last thing that a wheeled robot can handle without jerk. On the other hand a legged mechanism works with similar efficiency in both smooth and rough surfaces. The ultimate goal of Robosoccer is to design a team of humanoids that can defeat human world cup champion team. Thus any technique involved in it must match a bi-pedal robot or at least a legged robot. The availability of the AIBO enabled us to choose it for experiment with the proposed method. The nose camera [4] of the AIBO is used in measuring the ball distance process for the goalkeeper. Table 3.1 provides technical details about the hardware.

4   Experiments, Result and Discussion

4.1   Experiments with RL

The following experiments were devised for the evaluation of the Q-learning method against the hand coding based logic.

One attacker experiment designed with 3x7state action table using penalty shots taken from different points. The following pictures elaborate three different points used by the attacker to shot.

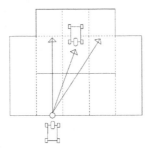

Fig. 1 Different one attacker layouts

The pictures printed in figure 1 shows one of the one attacker scenarios used to shot at the goal.   The space quantization discloses that the area in between the attacker and goal keeper was divided into six different sectors. The goalkeeper does not consider the area within the goal. Depending upon the space quantization 7 different basic situations was devised. They are named with s_1 to s_7 and listed below.

s_1 →Ball is at far left, s_2 →Ball is at far right, s_3 →Ball is at far middle, s_4 →Ball is right in front and heading straight towards goalkeeper, s_5 →Ball is located at close left corner, s_6 →Ball is located at close left corner, s_7 →Ball is in close front region,

Three basic actions are defined as:

$a_1 \rightarrow$ Go Right, $a_2 \rightarrow$ Stay where you are, $a_3 \rightarrow$ Go left

The following equation was used to evaluate the learning through the training process.

$$V(S_t) \leftarrow V(S_t) + \alpha * [R_{t+1} - \gamma * V(S_{t+1}) - V(S_t)]$$

$R_{final} = 100 = R_{t+1}$ (It becomes the final reward when reaches the terminal state)

$\alpha = 0.9$, $\gamma = 0.8$

These parameters were used to complete the training epochs and the knowledge base obtained from one attacker experiment was applied to the two attacker experiment described using a figure displayed in figure 2. There are three other different scenarios in terms of the position of the attackers exist in the two attacker mode.

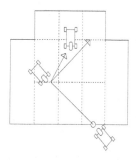

Fig. 2 Multi attacker layouts: Attacker arrangement 1

As displayed in the Figure 2 the attackers are placed in different positions in four different combinations. Both the one attacker and two attacker scenario was used to generate test results for the experiments. An OOV situation was also tested using the scenario shown in Figure 3.

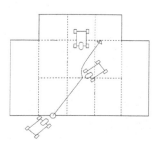

Fig. 3 Layout for an OOV situation

4.2 Ball Distance Experiment with IR Sensor and Nose Camera

The ball distance experiment setup is shown in Figure 4.

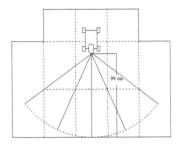

Fig. 4 Layout for ball distance measurement experiment

Two sets of data was taken using near and far IR sensors with the given set up as shown in Figure 5.

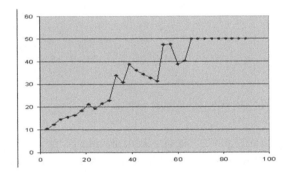

Fig. 5 Near Sensor data

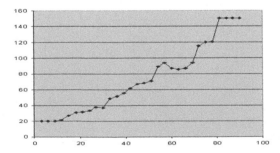

Fig. 6 Far sensor data

The ball.ratio data taken with nose camera is printed in Figure 7. Different co-lours on the graph signify that different reading was taken at a given point and so the average value was taken into consideration. The spline feet method was used to measure the distance in between two adjacent points situated 3 centimetres apart.

Te data table is used to create the value of m mentioned in Equation 2 for each interval separately and using the obtained ball.ratio value, the corresponding dis-tance was calculated with a greater accuracy than that of IR sensors.

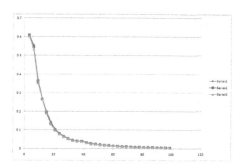

Fig. 7 Nose camera data

4.3 Benchmark

The bench mark database was made using both the single and double attacker sce-nario using Upenn'03 code base for the RL experiment and IR sensory data was considered as a benchmark of the ball distance measurement experiment.

4.4 Results

The following sections describe the results of the proposed and the base line me-thod.

4.4.1 Ball Distance Measurement Result with the IR Sensor and the Nose Camera

The IR sensors yielded 34.7 percent accuracy during the experiment where as the camera method provided 97% accuracy which is much higher than the last me-thod. More over the camera can detect few colours using proper filter. As a result the camera method was extensively used to detect and monitor the movement of the ball during the experiments.

4.4.2 One Attacker Experimental Result with 3x7 State x Action Table and Bench Mark

Table 4.3 was obtained as a converged matrix of the Q-learning process as printed below.

Table 2

	ACTION			
		a_1	a_2	a_3
S T A T E	s_1	0	64	0
	s_2	0	64	0
	s_3	0	64	0
	s_4	0	100	0
	s_5	0	0	80
	s_6	80	0	0
	s_7	0	80	0

The AIBO robots were made to perform using the table stated above. Also the Upenn'03 code base has been used for the same to create a base line to compare the efficiency of the training data using Q-learning. The benchmark showed that overall success of 79.9% achieved by Upenn'03. It was noticed that the Q-learning experiment with 3x7 state x action table yielded a success rate of 80.7 percent. This success rate is similar to the result obtained from the experiment conducted with Upenn'03 code base. However, unlike the hand-coding, the agent started from zero knowledge and ended up with virtually 100 percent efficiency. The limitation of physical manoeuvre stops the robot to achieve the same physically.

4.4.3 Results from the Two Attacker Experiment Using the Knowledge Base Obtained from One Attacker Training Data

The results of the four cases are described in the next table using both Q-learning and Upenn'03 code base. Altogether 20 shots were used to evaluate each and every separate attacker formations as one of them is displayed in Figure 4.2. The number of goals saved by the Q-learning data base and the hand-coding logic are given in the Table 3.

The experimental results showed that the agent has learned the knowledge of goalkeeping using proposed Q-learning from zero experience and without human interference. The results printed in Table 4.2 reveals that the Q-learning agent achieved an overall 77.5% success in goalkeeping alongside the Upenn'03 code that produced 72.5% efficiency.

The success rate of proposed Q-learning in the one attacker problem was better than that of Upenn'03 code base. Moreover the same database was applied to the two attacker problem; Q-learning method produced better efficiency than hand coding process as shown in OOV (Out of vocabulary) situation. The base line code hardwires the intelligence into the agent with fixed final decision at a given situation. However, the endless number of real life situation makes it critical to intercept and write a knowledge base for each and every case separately. As a result an unknown situation may results in an undesirable action. The same problem in the OOV situation given in Figure 4.3. The static rule based Upenn'03 code

Table 3

	Success rate achieved by Upenn'03 code base	Success rate achieved by Q-learning
Attacker arrangement 1	15	15
Attacker arrangement 2	16	14
Attacker arrangement 3	14	15
Attacker arrangement 4	17	14

went haywire due to certain change in the direction of the balls. On the other hand the proposed Q-learning helps the agent to map an optimum situation according to the present situation. Using Q-learning the agent took right decision properly. So, it can be concluded here that the agent learns to take the right action at the right moment using semi-supervised nature of learning. During the test process few shots took unexpected turns in front of the goal which was similar to the given OOV for the Upenn'03 method. However, the QLSF handled the situation properly and AIBO was able to take right decision at the right moment. The experimental results were carried out successfully using the one attacker problem. To prove the extent of the new method the knowledge base, acquired against one attacker was tested against two attackers as well. It was found that proposed Q-learning allowed the agent to work satisfactorily in a novel situation. To support this idea, it should be mentioned that AIBO tracked and followed each and every incoming shots properly.

5 Conclusion

The ultimate goal of Roboosccer is to prepare a team of humanoids to defeat human world cup champions [19]. It demands a robot (humanoid) to be equipped with human like manoeuvrability, image and pattern recognition systems and thinking and decision making capacity. This paper investigates if reinforcement learning (RL) as a form of semi-supervised learning can address the goal keeper's decision making process inside one and two attacker scenario in Robosoccer in conjunction with the spline fit method using the nose camera. We proposed here a Q-learning technique with spline fit (QLSF) that defines the situation using space quantization and the existence of the ball within it using the distance and direction and fits the situation to the available actions according to the Q-learning table. The Q-learning table is built from the scratch using the training epochs and back propagation of the reward after the Q-learning formulation. The novelty of our approach is that it can train the agent using certain degree of freedom in terms of action selection policy that provides a balance to exploration versus exploitation problem without any human supervision as used in the rule-base system. On the other hand the nose camera method provides a highly efficient and noise free location compare to the noisy and inaccurate IR sensor method. The performance of Q-learning was compared with a baseline created by Upenn'03 hand-coding logic using the goalkeeping problem and the nose camera method against the derived data sheet using IR sensors only. Analytically the base line method provides an input versus output table whereas the Q-learning produces a dynamic mapping to the intercepted state and a set of actions using state x action table where as the IR sensor mix up reading taken from ball and other objects in the vicinity. The result clearly showed that the proposed method can successfully handle an OOV situation and the camera produced a far better result than that found with the IR sensors. Further development possibilities using higher dimensional Q-learning features and a high speed and high resolution camera that come to the next stage of this research.

References

1. Mackworth, A.: Computer Vision: System, Theory, and Applications. World Scientific Press, Singapore (1993)
2. Sony, Qrio (2004), http://www.sonyaibo.net/aboutqrio.htm
3. Honda, Asimo (2008), http://asimo.honda.com/
4. Sony, Aibo (2006), http://support.sony-europe.com/aibo
5. Kokoro, Actroidder (2008),
   http://www.kokoro-dreams.co.jp/english/robot/
   act/index.html
6. Operating system documentation project (2003),
   http://www.operating-system.org
7. Currie, A.: The history of robotics
8. Fujita, M., Kitano, H.: Development of autonomous robot quadruped robot for robot entertainment. In: Autonomous Agents, pp. 7–18. Springer, Heidelberg (1998)
9. Coradeschi, S., Karlsson, L., Stone, P., Balch, T., Kraetzschmar, G., Asad, M., Veloso, M.: Overview of robocup 1999. Springer, Heidelberg (2000)
10. Stone, P., Balch, T., Kraetzschmar, G.K. (eds.): RoboCup 2000. LNCS (LNAI), vol. 2019. Springer, Heidelberg (2001)
11. Birk, A., Coradeschi, S., Tadokoro, S. (eds.): RoboCup 2001. LNCS (LNAI), vol. 2377. Springer, Heidelberg (2002)
12. Sammut, C., Uther, W., Hengst, B.: "runswift 2003", school of Computer Science and Engineering University of New South Wales and National ICT Australia (2003)
13. Kleiner, A.: Rescue simulation project (December 14, 2008),
    http://kaspar.informatik.uni-freiburg.de/rcr2005/
14. Niranjan, M., Rummery, G.A.: On-line q-learning using connectionist systems, Ph.D. dissertation, Cambridge University Engineering Department (1994)
15. Sutton, R.S., Barto, A.G.: Reinforcement Learning: An Introduction. The MIT Press, Cambridge (1998)

16. Teknomo, K.: Q learning numerical example (2006),
    `http://people.revoledu.com/kardi/tutorial/`
    `ReinforcementLearning/Q-Learning-Example.htm`
17. Negnevitsky, M.: Artificial Intelligence: Aguide to intelligent system, ch. 2. Pearson
    education, London (2002) ISBN 0-201-71159-1
18. Mukherjee, S., Yearwood, J., Vamplew, P.: Applying Reinforcement Learning in play-
    ing Robosoccer using the AIBO. In: GSITMS, University of Ballarat, Victoria,
    Australia (2010)
19. Robocup, Humanoid league 2006, `http://www.humanoidsoccer.org/`
20. URBI AIBO home page,
    `http://www.urbiforge.org/index.php/Robots/Aibo`

Author Index

.